INSTANT RECOMMENDATION LETTER KIT

INSTANT RECOMMENDATION LETTER KIT

HOW TO WRITE WINNING LETTERS OF RECOMMENDATION
FOURTH EDITION

By Shaun Fawcett, M.B.A.

Recommendation Letter Style Guide With 89 Fully-Formatted Real-Life Templates

RECOMMENDATION LETTERS: employment, college…
REFERENCE LETTERS: employment, character, general…
COMMENDATION LETTERS: employment, community service…
PERFORMANCE EVALUATION LETTERS: college, university…
SPECIAL BONUS CHAPTER: College Admission Essays

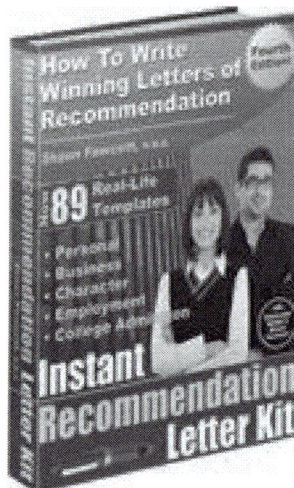

National Library of Canada Cataloguing in Publication Data

Fawcett, Shaun, 1949-
Instant recommendation letter kit [electronic resource] : how
to write winning letters of recommendation / Shaun Fawcett. – 4th ed.

Includes bibliographical references and index.

ISBN 978-0-9812898-4-7

1. Employment references. 2. Letter writing. I. Title.
PE1483.F378 2012   808'.06665   C2012-901031-0

Final Draft Publications

911-400 rue de l'Inspecteur

Montreal QC, Canada H3C 4A8

http://www.WritingHelpTools.com

TABLE OF CONTENTS

PREFACE TO SPECIAL PRINT EDITION I

INTRODUCTION 1
   REVISED EDITION NOTES 1
   BACKGROUND 2
   WHO THIS GUIDE IS FOR 4
   WHAT THIS WRITING KIT COVERS 5
   A DEFINITIVE WRITING RESOURCE 8
   YES, I DO REPEAT MYSELF 9
   DON'T SKIP THE FIRST 60 PAGES! 9

RECOMMENDATION LETTERS VS REFERENCE LETTERS 11
   THE CONFUSION 11
   BOTTOM LINE STRATEGY 12

RECOMMENDATION LETTERS EXPLAINED 15
   RECOMMENDATION LETTERS DEMYSTIFIED 15

RECOMMENDATION LETTER GUIDELINES 19
   RECOMMENDATION WRITING STRATEGIES 19

WRITING RECOMMENDATION LETTERS 23
   RECOMMENDATION WRITING TIPS AND POINTERS 23
   LETTER FORMATTING GUIDELINES 27
   HOW TO KEEP YOUR LETTER ON ONE PAGE 29

REAL-LIFE TEMPLATES FOR SUCCESS 33
   REAL-LIFE TEMPLATES FOR BETTER LETTERS 33
   ADVANTAGES OF REAL-LIFE TEMPLATES 36
   WORKING WITH REAL-LIFE TEMPLATES 37
   WARNING: BEWARE OF AUTOMATIC LETTER GENERATORS 39

THE TEMPLATE ADAPTATION METHOD 41
   USING THE TEMPLATE ADAPTATION METHOD 41

BENEFITS OF TEMPLATE ADAPTATION METHOD 50

RECOMMENDATION POWER PHRASES 51
OPENING STATEMENTS 51
ASSESSMENT STATEMENTS 55
CLOSING STATEMENTS 59

RECOMMENDATION LETTER TEMPLATES 63
TYPES OF RECOMMENDATION LETTERS 63
WRITING RECOMMENDATION LETTERS 64
EMPLOYMENT-RELATED RECOMMENDATION LETTERS 64
   Drafting Tips – Employment-Related Letters 64
SAMPLE TEMPLATES - EMPLOYMENT RECOMMENDATION LETTERS 67
   Recommendation 1: Employment – Marketing 68
   Recommendation 2: Employment – Supervisor 69
   Recommendation 3: Employment - Physician 70
   Recommendation 4: Employment - Occupational Therapy 71
   Recommendation 5: Employment – Graphic Artist 73
   Recommendation 6: Employment – Project Support 74
   Recommendation 7: Employment – Co-Worker 75
   Recommendation 8: Employment - Short-Term 76
   Recommendation 9: Employment – Student 77
   Recommendation 10: Employment – Manager 78
   Recommendation 11: Employment – Sales (Fashion) 79
   Recommendation 12: Employment - Technical Internship 80
   Recommendation 13: Employment - Refusal to Write 81
COLLEGE-RELATED RECOMMENDATION LETTERS 82
   Drafting Tips – College-Related Recommendation Letters 82
SAMPLE TEMPLATES - COLLEGE-RELATED RECOMMENDATION LETTERS 87
   Recommendation 1: College – Undergraduate 88
   Recommendation 2: College - Mature Student 89
   Recommendation 3: College – Undergraduate - Job 90
   Recommendation 4: College - Graduate - Psychology 91
   Recommendation 5: College – Graduate - Business 93
   Recommendation 6: College – Graduate - Stanford 95
   Recommendation 7: College - MBA Program - Haas 97

Recommendation 8: College - Graduate IT Program 99

Recommendation 9: College – Master of Journalism 101

Recommendation 10: College - MBA Program (UK) 102

Recommendation 11: College - Ph.D. Candidate 104

Recommendation 12: College – Special Award 105

Recommendation 13: College - Scholarship Applicant 106

Recommendation 14: To Recommender – Dental 108

SAMPLE TEMPLATES – GENERAL RECOMMENDATION LETTERS 109

Recommendation 1: General – Court Appointment 110

Recommendation 2: General – Community Work 111

Recommendation 3: General – Corporate 112

Recommendation 4: General - Special Award 113

Recommendation 5: General - Travel Visa Application 114

REFERENCE LETTER TEMPLATES **115**

TYPES OF REFERENCE LETTERS 115

WRITING REFERENCE LETTERS 116

Drafting Tips – Reference Letters 117

SAMPLE TEMPLATES – EMPLOYMENT-RELATED REFERENCE LETTERS 120

Reference 1: Employment – Sales Position (insurance) 121

Reference 2: Employment – Sales Position (pharma) 122

Reference 3: Employment – Sales Manager (IT) 123

Reference 4: Employment - MIS Manager 124

Reference 5: Employment - Student 125

Reference 6: Employment - Marketing 126

Reference 7: Employment – Plant Engineer 127

Reference 8: Employment - Software Support 128

Reference 9: Employment – IT Technical Specialist 129

Reference 10: Employment - Financial Specialist 130

Reference 11: Employment - Financial Controller 131

Reference 12: Employment - Customer Services 132

Reference 13: Employment – Research Specialist 133

Reference 14: Employment - Professional Consultant 134

Reference 15: Employment - Professional Services 135

Reference 16: Employment – Security Services 136

Reference 17: Employment - High School Teacher 137

Reference 18: Employment - Student Teacher (favorable) .... _138_

Reference 19: Employment - Student Teacher (neutral) .... _139_

Reference 20: Employment - Explain Departure .... _140_

SAMPLE TEMPLATES – COLLEGE-RELATED REFERENCE LETTERS 141

Reference 1: College - Graduate Studies (Education) .... _142_

Reference 2: College - Undergraduate Scholarship .... _143_

Reference 3: College - Teaching Award .... _144_

Reference 4: College – Exchange Program .... _145_

Reference 5: College – Business Undergrad Program .... _146_

Reference 6: College – Med. Residency: Emergency .... _147_

Reference 7: College – Med. Residency: Plastics .... _148_

Reference 8: College – Med. Residency: Neurosurgery .... _150_

SAMPLE TEMPLATES – CHARACTER-RELATED REFERENCE LETTERS 152

Reference 1: Character - Friend .... _153_

Reference 2: Character - Colleague .... _154_

Reference 3: Character – Domestic Services .... _155_

Reference 4: Character – Long-Time Client .... _156_

Reference 5: Character - Community Colleague .... _157_

Reference 6: Character – Committee Member .... _158_

Reference 7: Character - Rehabilitated Ex-Convict .... _159_

SAMPLE TEMPLATES – GENERAL REFERENCE LETTERS 160

Reference 1: General – Business .... _161_

Reference 2: General - Counseling Service .... _162_

Reference 3: General - Bank Reference .... _163_

Reference 4: General - Customer Credit Reference .... _164_

Reference 5: General - Landlord Re: Pet Owner .... _165_

Reference 6: General - Tenant .... _166_

COMMENDATION LETTERS **167**

DRAFTING TIPS – COMMENDATION LETTERS 167

SAMPLE TEMPLATES - COMMENDATION LETTERS 170

Commendation 1: Corporate - Employee .... _171_

Commendation 2: Customer Service - Employee .... _172_

Commendation 3: Teacher - From Parent .... _173_

Commendation 4: Award Nomination - Corporate .... _174_

Commendation 5: Community Service - Volunteer .... _175_

PERFORMANCE EVALUATION LETTERS **177**

  DRAFTING TIPS – EVALUATION LETTERS 177

  SAMPLE TEMPLATES - EVALUATION LETTERS 180

    Evaluation 1: Teaching - Satisfactory.. 181

    Evaluation 2: Teaching - Excellent 183

    Evaluation 3: Teaching - Borderline 186

    Evaluation 4: Teaching - Unsatisfactory... 189

    Evaluation 5: Teaching - Request for letter 191

    Evaluation 6: General Surgery - Residency... 192

COLLEGE ADMISSION ESSAYS **195**

  ADMISSION ESSAY TERMINOLOGY AND REQUIREMENTS...................195

  ADMISSION ESSAY REVIEW PROCESS 196

  WHAT THEY'RE LOOKING FOR 197

  DRAFTING TIPS – COLLEGE ADMISSION ESSAYS 198

  SAMPLE TEMPLATES – COLLEGE ADMISSION ESSAYS 205

    Admission Essay 1: Life-Changing Experiences....................................... 206

    Admission Essay 2: Travel and Cultural Diversity.................................... 208

    Admission Essay 3: Targeted University.. 210

    Admission Essay 4: Career and Personal Goals....................................... 212

    Admission Essay 5: Social Issues and Concerns....................................... 214

  COMMERCIAL WRITING SERVICES – COLLEGE ADMISSION ESSAYS 216

ONLINE RESOURCE LINKS **217**

  TOP 15 RECOMMENDATION LETTER SITES 218

  GENERAL WRITING REFERENCES 220

    Writing Style References... 220

    Writing Tools 222

INDEX **225**

Special Note Re: Hyperlinks

Because this book was first created and published online as a digital download eBook with "live" clickable hyperlinks throughout, those same links have been displayed in their entirety in this printed version for your information and use. Even though the links are not clickable in this paperback version of the book, they still provide pertinent website URL addresses that you can type into your Internet browser should you want to explore online for further information.

The hyperlinks are easily identifiable because they are <u>underlined</u> just as they would be on a website or in an eBook.

Dedication

The Fourth Edition of this book is dedicated to the thousands of people from all over the world who purchased the earlier editions of this book/ebook, thus helping it to become the world's leading English language resource on how to write ALL types of recommendation and reference letters. Thanks for that!

PREFACE TO SPECIAL PRINT EDITION

Instant Recommendation Letter Kit was initially conceived and written as an online eBook. Since 2002, the first three editions of the eBook have been available to purchase from a dedicated website:
http://www.instantrecommendationletterkit.com

From the moment that website went "live" the eBook version sold very well online because buyers were able to purchase and download it immediately to their computer hard drive.

Nevertheless, I realized from the beginning that no matter how much I publicized that website, many people in need of recommendation letter writing help and information would never become aware of it. I also knew that a certain group of people would prefer a standard paper book over the eBook version.

So I decided to also use alternative, more conventional distribution channels.

Consequently, this particular version is a special one that is being made available through more traditional book marketing channels. So, if you are reading this paperback version, it should be because you purchased it through a large traditional book distribution company such as amazon.com, barnesandnoble.com, or other similar standard book retailers.

Get Your Downloadable Templates

There is one key difference between this book that you purchased through a book distributor or retailer and that which you would have received had you purchased and downloaded the eBook version from my website.

Website purchasers can instantly download the Bonus Templates (MS-Word format) straight to their computer at the time of purchase. However, For technical reasons, that is not possible when purchasing the book through distributors such as amazon.com and conventional bookstores.

Nevertheless, since the templates are an integral part of the Kit, I wanted to make sure that book owners also have access to them.

So, to receive your Free copy of the Bonus Templates file, please send an e-mail to the following special address:
templates47r@instantrecommendationletterkit.com

In that e-mail you MUST include the following information: your full name, full telephone number, your primary e-mail address, and the date of purchase. Also indicate which business you purchased the Kit from, and include the Order Number or Transaction Number that you were given at the time of purchase.

Once I have received your e-mail and verified your purchase, I will send you the Bonus Templates as an attachment to an e-mail. You should receive it within a few hours of making your request, 24 hours at the most.

This Book Has Two Parts

As explained above, when the eBook version is sold from my website it comes in two downloadable parts: 1. The Main Writing Kit eBook (pdf), and 2. The Bonus Templates (MS-Word), as follows:

1. **Instant Recommendation Letter Kit (244 page pdf eBook)**

2. **Bonus Templates (157 page MS-Word report)**

Again, Part 2, the downloadable Bonus Templates file, is NOT included with this paperback version of the eBook, but it may be ordered for Free by e-mail at the following address:
templates47r@instantrecommendationletterkit.com

Remember, when requesting your downloadable Bonus Templates make sure that you include the following information in your e-mail: *your full name, full telephone number, your primary e-mail address, and the date of purchase. Also indicate which business you purchased the Kit from, and include the Order Number or Transaction Number you were given at the time of purchase.*

INTRODUCTION

REVISED EDITION NOTES

The first version of *Instant Recommendation Letter Kit* was published both as an eBook and a paperback back in 2002. That was followed by revised editions in both 2005 and 2008. This **Fourth Edition** is also being released as both an ebook and a paperback, in early 2012.

All of these versions have contained the same "core" how-to content; the major change for each edition being the addition of more real-life templates.

Sales of both the book and ebook have been excellent since their initial release. In fact, this title quickly became my best selling "how-to" book and remains so until today. As far as I know, it is still the ONLY book/ebook in existence that deals with writing ALL types of recommendation letters.

A few websites offer specialized types of recommendation letters. These are essentially of the "one size fits all" copy and paste variety. No other book/ebook that I know of deals with them ALL: job/employment, college/university admission, character, community service, and performance evaluation.

The primary reason why I am releasing this **Fourth Edition** is so that I can share additional recommendation letter and reference templates that I have developed over the past couple of years. Also, in the intervening period I have learned a number of additional facts on the subject of writing recommendation letters that I would like to pass on to my readers.

The main ADDITIONS incorporated into the *Fourth Edition* are as follows:

- **Ten (10) more NEW professionally written full-length recommendation and reference letter templates** based on actual "real life" situations; **making a total of 89 downloadable templates!**

- **Reorganized and updated material explaining the important differences**, and similarities, between recommendation letters and reference letters.

- **A revised and expanded section on how to use the Template Adaptation Method™** to quickly and easily create recommendation and reference letters using the Kit's real-life templates.

- **An updated list of the "Top 15 Recommendation Letter Writing Sites"** to reflect the latest content on the Web.

I have no doubts whatsoever that this *Fourth Edition* of *Instant Recommendation Letter Kit* is the most comprehensive resource that you will find ANYWHERE on the subject of writing ALL types of recommendation letters.

BACKGROUND

The original version of *Instant Recommendation Letter Kit* was the second in a series of practical "writing kits" that I developed based on the use of "real-life" sample templates that people can download into their word processing program and work with. Since then, my series of writing toolkits has grown to include:

Instant Business Letter Kit
How To Write Business Letters That Get The Job Done
http://instantbusinessletterkit.com

Instant Resignation Letter Kit
How To Write A Super Resignation Letter And Move On With Class
http://instantresignationletterkit.com

Instant Letter Writing Kit
How To Write Every Kind Of Letter Like A Pro
http://instantletterwritingkit.com

Instant College Admission Essay Kit
How To Write A Personal Statement Essay That Will Get You In
http://instantcollegeadmissionessay.com

Instant Home Writing Kit
Hot To Save Money, Time and Effort and Simplify Everyday Writing Tasks
http://instanthomewritingkit.com

In addition to the above, I have written a couple of books that provide people with the step-by-step essentials for creating, publishing and marketing their own books/ebooks:

Instant Book Writing Kit
How To Write, Publish and Market Your Own Money-Making Book/eBook Online
http://instantbookwritingkit.com

How To Write A How-To Book (or eBook)
Make Money Writing About Your Favorite Hobby, Interest or Activity
http://howtowritehowto.com

Instant Recommendation Letter Kit was my first writing kit to focus on one specific type of letter writing requirement. It concentrates on the development and writing of ALL types of "recommendation letters", whether for business, personal, or college admission purposes.

In fact, there are five (5) primary recommendation letter "types", as follows:

- recommendation letters – job related

- letters of recommendation – college/university admission related

- reference letters – job and community-related

- commendation letters – employment and community related

- performance evaluation letters – employment related

The idea for this particular book grew directly out of visitor reaction to my first writing help Web site:
http://www.writinghelp-central.com.

After monitoring hundreds of thousands of visitors to that website during its first year of operation, it became clear that the large majority of people who arrived there were looking for letter-writing information and samples specifically related to the keywords "letter of recommendation" and "recommendation letter." Clearly, there was a demand for information on how to write recommendation letters.

As a result, I concluded that there was a definite need for a practical quick reference "recommendation letter kit" with "real-life" sample templates. Based on sales since then, my conclusion at that time has proven correct.

Nowadays, of the more than 10,000 unique visitors to that website each and every day, on average, almost 50% of them are looking for information and samples on the subject of writing some type of recommendation letter.

Accordingly, _Instant Recommendation Letter Kit_ has been designed as a quick-reference home and home-business writing guide, focused on simplifying the often difficult and delicate task of writing the various types of recommendation letters. It is a practical hands-on "toolkit" that people can use whenever they have to write a recommendation letter of any kind.

Using the 89 "real-life templates" included with the Kit, you will never again have to start a letter of recommendation from a blank page or screen. You can work directly from the real-life templates that you can download straight into your word-processing program. (MS-Word compatible).

WHO THIS GUIDE IS FOR

This Kit was written to help ANYONE who needs to write ANY type of recommendation letter for business or personal reasons.

Since the first version of this book went on sale, thousands of copies have been sold to a wide variety of businesses, institutions and individuals including:

- high school teachers and administrators
- college/university professors
- college/university students
- professionals (all disciplines)
- business owners/operators
- legal and employment firms
- high school and college students
- government employees/administrators
- military personnel/administrators

- NGO administrators
- community associations
- private citizens/individuals

As stated earlier, this book contains multiple templates covering just about every kind of recommendation or reference letter situation imaginable.

Bottom Line: Whether you're a business person, an educator, a professional, a private citizen, a student, or an employee of a government or corporation, this Kit will answer your recommendation letter-writing needs in over 99% of all cases.

In fact, if you find a case in which one of the sample templates in the Kit cannot be easily adapted to your need, let me know. If I agree that there is no template in the Kit that can be used for your letter, **I'll write the letter for you!**

Now, how's that for a guarantee!

WHAT THIS WRITING KIT COVERS

This letter-writing toolkit is much more than just a bunch of templates quickly thrown together, as is the case with the majority of manuals of this type.

The Kit is a complete and comprehensive letter-writing style guide that contains more than 85 fully-formatted real-life recommendation letter templates. It is comprised of **two major components.**

Component number one is a 60-page recommendation letter writing "how-to" style guide combined with the fully-formatted real-life sample letters.

Component number two is a set of more than 85 fully-formatted real-life letter templates that can be downloaded straight into a standard word processing program such as MS-Word. Using these downloadable templates, you will never have to start your letters from a blank page again.

The Style Guide

The style guide part of the Kit contains **almost 60 pages of letter-writing "how-to" information** including letter-writing tips, strategies and information as well as letter formatting guidelines for writing ALL types of recommendation letters.

Over 85 <u>fully-formatted</u> real-life sample letters are included to graphically demonstrate how to write each letter-type covered. These are REAL letters, written by professionals -- **NOT** your typical fill-in-the-blank, cut-and-paste jobs found at a few sites on the Internet.

This manual covers ALL known types of letters of recommendation. These letters are most often employment-related or education-related. Following is a list of the most common types of letters that fall into the broad "letters of reference" category.

Recommendation Letters

Also known as letters of recommendation, the majority of recommendation letters are related to employment, and admission to college and graduate school.

Reference Letters

Often confused with their closely-related cousins, recommendation letters; letters of reference are usually employment-related, and character-related.

Commendation Letters

Letters of commendation are most often written in work-related or community service situations.

Performance Evaluation Letters

Performance evaluation letters are usually written in employment and school related situations.

This manual includes valuable tips, pointers and information to help you with the development and writing of just about any kind of letter of recommendation.

The Templates

For each of the **85 plus sample letters, a fully formatted real-life template is provided in a form that can be downloaded straight into your word processor** (MS-Word format). The templates have been divided into nine (9) groups as follows:

1. Recommendation Letters – Employment-related
2. Recommendation Letters – College-related
3. Recommendation Letters - Commendation Letters
4. Recommendation Letters - Performance Evaluation Letters
5. Recommendation Letters – General purpose
6. Reference Letters – Character references
7. Reference Letters – College references
8. Reference Letters – Employment references
9. Reference Letters – General references

Important Note:

People who buy this Kit directly from my website receive the templates file as a direct download at time of purchase. Those who buy the Kit from retailers such as amazon.com and others have the option to request that the templates file be sent to them by e-mail. The information for doing this is available in the Special Preface section at the beginning of paperback versions of this book.

Recommendation Writing Resources

In addition to the above, the Kit also includes other letter writing resources:

1. **Top 15 Web Sites** - My own exclusive fully-researched list of the "Top 15 Online Recommendation Letter Writing Sites".

2. **Top Writing References** - My personal list of recommended writing reference texts.

Instant Recommendation Letter Kit also includes a detailed Table of Contents and a Keyword Index for easy reference.

Writing College Admission Essays BONUS!

In addition to the various types of letters of recommendation a **20-page Bonus Chapter** has also been included on how to write an admission essay for college or university. If one needs a recommendation letter for college or university, one will also need to know how to write admission essays. (See page 195).

I have included this particular subject in a bonus chapter because I know that a large percentage of people who buy this Kit are graduate and undergraduate students who are in need of admission essay help and advice.

This bonus chapter includes five (5) fully-formatted real-life college admission essays that you can download and adapt to your situation.

A DEFINITIVE WRITING RESOURCE

While initially researching this topic I was amazed to discover all of the confusion surrounding the seemingly simple term "recommendation letter". I was even more amazed when I started looking around for a definitive and complete resource book on the recommendation letter in all of its various forms.

Well, since I did my first research, things have not changed much!

As far as I know, a comprehensive guide to writing ALL types of recommendation letters did not exist until I published the first version of _Instant Recommendation Letter Kit_ back in 2002.

Now, with this **_Fourth Edition_** it remains the ONLY complete and definitive guide on writing ALL types of recommendation letters. Using this revised edition as a guide, anyone needing to write any type of "recommendation letter" will find what they need somewhere in this guide.

However, nobody is perfect. If you find a true recommendation letter situation that is not covered in this guide, please let me know and I will make sure that the next version includes it. Just drop me a quick e-mail at:
mailto:info@instantrecommendationletterkit.com

YES, I DO REPEAT MYSELF

I'll tell you right now that there are a few places in this book where I repeat myself on purpose. When I have done so, those repetitions are deliberate and for good reason. Here's why…

Since this is a "how-to" reference guide, first and foremost, I know that some people will only refer to the sections/pages that apply directly to their immediate need. So, in a few places I have repeated what I consider to be critical information in order to let a particular chapter "stand alone" for those people who won't take the time to read the other sections of the book.

DON'T SKIP THE FIRST 60 PAGES!

I know from experience that the initial inclination of many people using this Kit to get help with their recommendation letter writing will be to jump straight into the sample templates without even pausing to look at the advice, tips and pointers at the beginning of this Kit.

If you skip the first 60 pages you will be missing lots of valuable/useful information!

The information included in the "style guide" part of the Kit is based on over 30 years of experience writing thousands of business and professional letters. The information contained in those first 60 pages will definitely help you improve the quality of your recommendation letters.

So, whatever you do, DO NOT MISS the section on "How To Use The Template Adaptation Method". (See page 41).

That technique alone is well worth the price of this Kit.

RECOMMENDATION LETTERS VS REFERENCE LETTERS

THE CONFUSION

I begin my article titled "Recommendation Letters Explained" on page 15 with the following sentence:

"There is a lot of confusion about recommendation letters".

Well, there still is folks! Yep, there remains plenty of confusion between the terms "recommendation letter" and "reference letter" in terms of exactly they mean and which is which.

It's confusing enough for the average person who only has to write one of these letters every once in a while. However, when you look into it in depth as I have and realize that many of the so-called "experts" don't even know the differences between a recommendation and a reference letter, the confusion is understandable.

I tried to demystify this confusion somewhat in the first version of this book by drawing a bit of a line between recommendation letters and reference letters. That helped, but it still left a gray area of overlap between the two.

Since then I've been watching the whole issue carefully and have conducted additional research, leading me to conclude the following:

- Many people and institutions use the term "recommendation letter" to mean BOTH recommendation letters AND reference letters (about 2/3).

- Many people and institutions use the term "reference letter" to mean BOTH reference letters AND recommendation letters (about 1/3).

- Both groups are talking about essentially the same type of letter in their minds and they don't seem to draw any distinction between the two.

- Thus, those who use the term "recommendation letter" ALWAYS use that term. Those who use the term "reference letter" ALWAYS use that term.

... and, "never the twain shall meet", as the saying goes.

So, for the purposes of this guide I am going to use the broad definitions that I originally stated in *Instant Recommendation Letter Kit,* as follows:

Recommendation Letters "Defined"

Originally, "recommendation letters" were letters that made a clear and specific "recommendation" about someone.

I define it as a "recommendation letter" when the letter is specifically requested about someone and is therefore addressed to a specific requestor.

Generally speaking, recommendation letters are almost always employment-related or college program admission/scholarship related.

Reference Letters "Defined"

Originally "reference letters" were general factual statements about someone and/or their situation that did not make any specific recommendation about the person.

I call it a "reference letter" if it is more general in nature, and is NOT addressed to a specific requestor. Typically "letters of reference" are addressed; "To Whom It May Concern" or "Dear Sir/Madam".

Reference letters include character reference letters and employment reference letters. Also, any kind of general "Dear Sir:" or "To Whom It May Concern" sort of letter such as one to a landlord or to a bank, is a reference letter.

BOTTOM LINE STRATEGY

In any case, somehow over the years the line between the two types of letters has become blurred and many people and institutions now use the two terms interchangeably. As far as those organizations are concerned both letters are exactly the same thing.

Accordingly, please note that many colleges and universities use the terms "reference letter" and/or "letter of reference" when referring to **exactly** the same thing as what many *other* colleges and universities call "recommendation letters" and/or "letters of recommendation".

So when dealing with these types of institutions make sure you use whichever term they use in the same way that they use it. Don't try to convert them to your terminology. That will just confuse things and you will pay the price.

One thing that I **can** tell you for certain is that the term "recommendation letters" is the more popular search term of the two when people are doing Internet searches.

For whatever reason, more people are looking for information on recommendation letters than on reference letters.

RECOMMENDATION LETTERS EXPLAINED

As I already mentioned, information about letters of recommendation and sample recommendation letters are by far the most often clicked letter-related links at my *www.writinghelp-central.com* website.

Typically, a recommendation letter conveys one person's view of the working performance and general workplace demeanor of a person who has worked under their direct supervision.

Recommendation letters are generally requested when the requestor is applying for a new job, or when they are trying to get accepted into college or university.

Of course, the usual expectation is that a recommendation or reference letter will be positive overall.

This section contains an expanded version of an article I wrote entitled *"Recommendation Letters Demystified"*. That article has been published in selected e-Zines and posted on various websites across the Internet.

RECOMMENDATION LETTERS DEMYSTIFIED

There is a lot of confusion about recommendation letters.

Recommendation letters are often referred to in a number of different ways including: letters of recommendation, reference letters, letters of reference, commendation letters, and sometimes even, performance evaluation letters.

This terminology can be quite confusing, especially when these terms are often used interchangeably, sometimes to mean the same thing, sometimes to mean something different.

Below are some definitions that should clear up any confusion, followed by some tips and strategies on how best to deal with recommendation letters.

Letters of Recommendation

Employment-Related

Also called a recommendation letter, it is an employment-related letter that is specifically requested by the person the letter is being written about. Such a letter is normally positive in nature, and written by someone who knows the subject well enough to comment on the skills, abilities, and specific work attributes of that person.

Typically, an employment-related letter of recommendation conveys one person's view of the work performance and general workplace demeanor of another person that has worked under their direct supervision. The requestor of the letter normally requests such a letter to use when applying for a promotion or a new job.

These letters are usually addressed to a specific person to whom the requestor has been asked to submit the letter.

College- and University-Related

Another situation where recommendation letters are a common requirement is for entry into undergraduate and graduate programs at a college or university. Graduate programs often require two or more letters of recommendation as part of the program admission requirements.

Normally, these college-related recommendation letters are written at the request of the program applicant by people who know them and are familiar with their academic career to-date, and their future education and career aspirations. These people could include: former teachers, community leaders, school faculty members, administrators, academic supervisors, and/or employers.

These letters are always addressed to a specific person and are normally included as part of the program admission application.

Letters of Reference

These are more general letters that are often requested by employees when they leave the employ of an organization. Normally factual in nature, they are usually addressed, "to whom it may concern" and provide basic information such as: work history, dates of employment, positions held, academic credentials, etc.

Employment Reference

Reference letters sometimes contain a general statement (as long as a positive one can be made), about the employee's work record with the company that they are leaving. Employees often submit these letters with job applications in the hope that the letter will reflect favorably on their chances for the new position.

Character References

Character reference letters are sometimes required by employers when hiring individuals to perform personal or residential services such as child care, domestic services, etc. These letters are usually drafted by a former employer and deal with such characteristics as honesty, dependability and work ethic/performance.

Commendation Letters

These are usually unsolicited letters, which typically commend an employee to their supervisor for something outstanding or noteworthy that the employee has done. Usually, the employee would have to do something "above and beyond" what is normally expected of them in their job to warrant such a letter.

Typically, these letters are written by co-workers or managers from another area of the organization who were suitably impressed while supervising the person on a short-term project.

Commendation letters are also used to nominate individuals for special awards of recognition for outstanding public service.

Performance Evaluation Letters

These are usually detailed assessments of an employee's work performance as part of an organization's regular employee review process. Typically, they are written by the employee's supervisor and are attached to the individual's performance appraisal and placed in their personnel file.

The format and structure for this type of letter is more often than not dictated by the employee performance evaluation system or process that is in-place wherever the subject of the letter is employed.

However, in the academic environment in North America there is often a requirement for a specific "performance evaluation letter" for the assessment of academic staff. A number of real-life templates of academic-related performance evaluation letters are included in the appropriate section of this guide.

In addition, the Online Resources Links chapter at the end of this guide contains link to various types of performance evaluation letters, from both the academic sector and other employment sectors.

RECOMMENDATION LETTER GUIDELINES

In addition to standard letter-writing protocol, there are a number of basic guidelines that cover most situations related to the writing of recommendation letters specifically. These are usually more "situational" than "how-to" in nature.

The "how-to's" of writing recommendation letters are covered a bit later on in the section titled "Writing Recommendation Letters" (See page 23).

These recommendation letter guidelines are important to both note and apply, since writing letters of recommendation is always a somewhat tricky and delicate matter. This is because they almost always affect the reputation and future of another person.

The following is another excerpt from an article I wrote entitled *"Recommendation Letters Demystified."* That article has been published in numerous e-Zines and posted on various websites across the Internet.

RECOMMENDATION WRITING STRATEGIES

The following strategies apply primarily to the writing of recommendation letters and reference letters as defined above.

Write It Only If You Want To

If you are asked by someone to write a recommendation letter about them, you don't have to say "yes" automatically. If it's someone you respect for their work, and you have mostly positive things to say, by all means write the letter.

There is no point saying "yes" and then writing a letter that says nothing good about the person, or worse still, concocting a misleading positive assessment of someone.

So, whatever you do, don't get sucked into writing a recommendation inappropriately out of feelings of guilt or obligation.

If You Must Refuse, Do So Right Up Front

On the other hand, if someone asks you to write a recommendation letter for them, and you know you'll be hard-pressed to keep it positive, say "no" right away.

There is no point in hesitating and leading the person on to believe that the answer might eventually be "yes".

A gentle but firm "no" will usually get the message across to the person. Explain that you don't think that you are the best (or most qualified) person to do it.

Suggest Someone Else

If you feel you should refuse, for whatever reason, it may be helpful for you to suggest someone else who you think might have a more positive and/or accurate assessment of the person.

That other person may be in a better position to do the assessment. Usually there are a number of possible candidates, and you may not actually be the best one.

In fact, I have seen a number of cases over the years in which people requesting recommendation letters have not requested the letter from the obvious or logical choice. This usually happens when the requestor doesn't like the person who is the obvious choice, and/or they are worried about what that person will say about them.

Write It As You See It

Writing a less than honest recommendation letter does no one a favor in the end. It is likely to backfire on you, the person being recommended, and the new employer.

Also, many employers and head-hunting agencies check references these days.

How would you like to be called up and have to mislead people due to questionable things you may have written in a reference letter?

Be Honest, Fair, and Balanced

Honesty is always the best policy when it comes to writing recommendation letters. At the same time, try to be fair and balanced in your approach.

If in your estimation, a person has five strengths and one glaring weakness, but that weakness really bothers you, make sure you don't over-emphasize the weak point in the letter based on your personal bias. Just mention it in passing as a weakness and then move on.

Balanced Is Best

An overall balanced approach is the best one for a letter of recommendation. Even if your letter generally raves about how excellent the person is, some balance on the other side of the ledger will make it more credible. After all, nobody's perfect.

There must be some area where the person being recommended needs to improve. A bit of constructive criticism never hurts and it will make your letter more objective in nature.

Bottom Line

The most important point to take from the above is that it is your choice as to whether, and how, you will write a recommendation or reference letter.

It's an important type of letter that will have a definite impact on the future of the person about which it is being written, so if you agree to write one make sure you give it your utmost attention and effort.

WRITING RECOMMENDATION LETTERS

Even though this guide is written specifically to cover the writing of letters of recommendation, many of the basic principles of letter writing in general still apply and should be observed.

Based on the feedback that I have been getting from visitors to the ***writinghelp-central.com*** website, general letter writing is definitely the area where most people are looking for help or guidance.

In fact, over 65% of the visitors to my site are seeking some sort of letter writing information or assistance. And, as I stated earlier, a large number of those are looking for information related to recommendation letters.

This section provides a brief review of the most important guidelines to follow when writing letters in general, and thus recommendation letters as well

The following is a revised and expanded version of an article I wrote entitled *"7 Essential Letter-Writing Strategies"*. That article has been widely published in eZines and posted on numerous websites across the Internet. I have adapted it here to focus on the writing of recommendation letters.

RECOMMENDATION WRITING TIPS AND POINTERS

Following are a few practical letter-writing tips adapted from the ***writinghelp-central.com*** website to help you when writing that next recommendation letter:

Keep It Short and To The Point

As a general rule, recommendation letters should be concise, factual, and focused. Try to never exceed one page or you will be in risk of losing your reader. A typical letter page will hold about 350 to 400 words.

However, there can be cases that will warrant two pages; NEVER go beyond two.

If you can't get your point across with that many words you probably haven't done enough preparatory work. If necessary, call the recipient on the phone to clarify any fuzzy points and then use the letter just to summarize the overall situation.

Focus On the Recipient's Needs

While writing the letter, focus on the information requirements of your audience, the intended addressee. If you can, in your "mind's eye", imagine the intended recipient seated across a desk or boardroom table from you while you are explaining to them, the subject of the letter.

What essential information does that person need to know through this communication? What will be their expectations when they open the letter? Have these all been addressed?

Use Simple and Appropriate Language

For clarity and precision, your letter should use simple straightforward language. Use short sentences and don't let paragraphs exceed three or four sentences. As much as possible, use language and terminology familiar to the intended recipient.

Don't use technical terms and acronyms without explaining them, unless you're certain the addressee is familiar with them.

Reread and Revise It

Do a first draft, and then carefully review and revise it. Put yourself in the place of the addressee.

Imagine yourself receiving the letter. How would you react to it? Would it answer all of your questions? Does it deal with all of the key issues? Are the language and tone appropriate?

Sometimes reading the letter "out loud" to one's self can help. When you actually "hear" the words, it is easy to tell if it "sounds" right, or not. I do this all the time and it really works!

Use Transition Words and Phrases

One method that I always use to help with the flow and sequencing of my text in letters is to employ "transition words" throughout.

These are great for allowing you to connect thoughts and create logical links/flow between sentences and paragraphs.

These transition words and phrases are usually inserted at the beginning of a sentence and they refer directly back to the previous sentence and/or paragraph without repeating the specific subject. They allow you to maintain a logical flow and make smooth transitions from one thought to the next.

Some typical transition words/phrases are: then, as a result, unlike, different, in spite of, next, in addition, like, the same, similar, for example, one such, for instance, accordingly, etc.

When using transition words/phrases, remember that they almost always ***refer back*** to the previous sentence or paragraph.

Eliminate Redundant Words and Phrases

There are certain words and phrases that one often sees in business correspondence that tend to make the language more complicated and cumbersome than necessary.

For example, some typical "redundant words/phrases" would include: "absolutely essential" instead of "essential", "actual experience" instead of "experience", "attached hereto" instead of "attached", "as a result of" instead of "because", "few in number" instead of "few", etc.

These are just a few examples. I'm sure you can think of others. Always look for redundant words/phrases when reviewing your final draft letter.

If a word or phrase doesn't add value and/or meaning, omit it.

Check Spelling and Grammar

A letter is a direct reflection of the person sending it, and by extension, the organization that person works for. When the final content of the letter is settled, make sure that you run it through a spelling and grammar checker.

To send a letter with obvious spelling and grammatical errors is sloppy and unprofessional. In such cases, the recipient can't really be blamed for seeing this as an indication as to how you (and/or your organization) might do most other things.

Spell-checkers are great, but they don't catch everything. For example, I often reverse the letters in certain words when typing quickly. i.e. "form" instead of "from." A typical spell-checker would say these are both valid. Some grammar checkers will flag it as "out of context", but you can't always count on that.

The only way to be sure in the end that everything is fine, is to have someone with good spelling and grammar skills do a final visual check.

Bottom Line

The above basic letter writing tips are mostly common sense. Nevertheless, **you would be amazed** at how often these very basic "rules of thumb" are not employed when people are writing letters.

I believe that observing the above letter writing guidelines are particularly important when writing a recommendation letter or reference letter. Failing to observe one or more of these points will not only reflect poorly on you as the writer, but may well have a negative impact on the person being recommended.

Accordingly, since a recommendation letter will almost always impact some aspect of the future of another person, be particularly careful to observe ALL of the foregoing pointers.

LETTER FORMATTING GUIDELINES

The following guidelines are adapted from a section of one of my other writing Kits that deals with business letters specifically:

http://www.instantbusinessletterkit.com

I include them here because they are also applicable to writing just about any professional-looking letter.

Again, these are recommended guidelines only. Nothing is cast in concrete. Although I wouldn't suggest you deviate from them unless you have good reason to do so. Vary them as you see fit, within reason, or as you may require in order to "squeeze" things onto one page. Remember, if you do "squeeze" things a bit, it shouldn't be easily detectable by the average reader.

In other words, it shouldn't *look* "squeezed". If it does, you will have to go to a second page (See next section for tips on "compressing" a letter onto one page).

- Top and bottom, and left and right page margins, can vary from 1 inch to 1.5 inches, but should be (or appear to be) the same all the way around.

- I recommend using a font style and size of Times Roman, 12 point. This combination has a businesslike appearance and is widely used.

- The main text of resignation letters should be single spaced, with double spacing between paragraphs.

- A typical one-page letter will have between four (4) and (5) paragraphs. Try to limit paragraphs to two or three sentences each in order to achieve a balanced look that is not too dense.

- Generally speaking, skip one (1) line horizontally between the different components of the letter (as per sample templates, later).

- At the top of the letter, try to skip between two (2) and three (3) lines between the Letterhead Block and the Recipient Address Block.

- In the Signature Block, leave three (3) to four (4) lines between the Closing Salutation (i.e. Sincerely), and the Signatory Name so that there is ample room for a normal hand-written signature.

- For a general idea of spacing, refer to the sample formats presented earlier. The most important thing is "relative" spacing. It is not hard to recognize when two items are too close together or too far apart.

- After you have finished the letter you may want to make some final adjustments to spacing in order to give it a "balanced" look.

In my professional opinion, the spacing in letters is not an "exact science" as it once was. What you're looking for is an uncluttered, balanced look that respects the basic format you are following. So, don't get too worked-up about absolute formatting perfection! Just make sure your letter looks good.

I've written thousands of business letters over the years and I can't recall one of them ever being ignored because the margins were 0.8 inch rather than 1.0 inch. If it's uncluttered and well-balanced in appearance (and most importantly, well-written!), it will be taken seriously as a professional business letter.

As far as punctuation and capitalization are concerned, please refer to the "real-life templates" provided later on to see specific examples in specific contexts.

You can also reference any one of the writing guides listed under "Writing References" in the Resources chapter at the end of this guide.

HOW TO KEEP YOUR LETTER ON ONE PAGE

There are a number of handy little tricks that I have learned over the years that will help "squeeze" a letter onto one page without it being noticed by the average reader. These tricks can be applied using any standard word processing program.

I'm not sure whether a purist at a secretarial school would approve of some of my methods since they may deviate from certain standards, but I have used them hundreds of times and nobody has ever been the wiser. The main point being that I was able to keep a letter on one page when the first version overflowed by a few lines onto a second page.

Below are my "page compression tips", in the order I suggest you use:

1. Move the left and right margins out about ¼ in. closer to the edge of the page.

2. Move the top and bottom margins out about ¼ in. closer to the edge of the page.

3. Edit out the one or two word "overflows". What I mean, is this: After the letter is drafted take a good look at each paragraph. See if there are any that have an ending sentence that "overflows" onto an additional line for the sake of one word. If so, make a minor edit or two in the paragraph that shortens it a little so that the last word or two will not overflow onto the following line. Using this method, you can often gain two or three extra lines in a one-page letter.

4. Adjust the line spacing on the page. You can gain considerable space on a page by adjusting the line spacing of the text. For example, if the default line spacing is set to "single" at 12 points try setting it to "exactly" at 12 points if your font size is 12. If that doesn't do it, try "exactly" at "11 pts". Often you have to experiment a bit with this one to get the look just right.

5. As a last resort, try reducing the font size by 1 point, say from 12 to 11.

6. If it still doesn't "fit", there's one final thing you can try if you're the author of the letter. Go back and edit it one more time. Look for redundant thoughts and phrases, or those that can be combined into one sentence rather than two. Is every

word and phrase absolutely essential to your message? You'll be amazed at the space savings that this process can result in.

As I said earlier, try the above methods in sequence, one-at-a-time, checking each time to see if your latest change has done the trick for you.

What happens if it still won't fit?

Now, if you've used all of the above tricks and you still can't get the letter to fit onto one page, it's time to admit that you've got a real two-pager. In which case, you should then think about "reversing" some or all of the compression tricks that you applied when you tried to "squeeze" the letter onto one page. Then, concentrate on making a balanced looking second page.

There's nothing worse looking than a letter with a one or two sentence second page! So in this case, you may want to actually "stretch" the letter out a bit.

The first thing I do in this instance is increase the line spacing and reduce the margins slightly so that there will be a decent sized overflow onto the second page.

So, try reversing steps 1, 2 and 4 above. Instead of decreasing the top, bottom and side margins on page one, try increasing them by ¼ in. all around. Then adjust the point size and see if that helps.

Again, I have used these little "compression" tricks thousands of times, and nobody has ever pulled out their ruler and chastised me for inaccuracy. The important thing is to end up with a professional-looking letter.

In fact, if you do a detailed check on the letter templates included later in this guide you will find that I have used one or more of the above tricks on many of them.

But, I'm not telling which ones!

Letter Formats Used In This Guide

This Kit includes more than 85 sample recommendation letter templates. As already discussed, there are two main page formats/layouts that can be used for almost all letters: Block Style and Semi-Block Style.

Since there is no international standard for formatting business letters I have decided to use my personal favorite letter format in all of the templates/samples throughout this document. It's called the ***Modified Semi-Block Style***. Essentially, it places all major blocks flush to the left-hand margin, except for the date and return address blocks at the top of the letter, which are placed flush to the right margin.

Most large organizations have a "corporate style manual" that specifies the standard formats for letters at that particular organization. If you don't have such a standard, simply copy the exact formats used in the templates included in this Kit. You can't go wrong with those.

The main thing is to choose one style that you like and stick with it.

Notes Re: Sample Letter Formats

- Although the sample templates in this guide are based on actual situations involving real people, identifying details have been altered to protect privacy.

- All of the sample templates in this guide have been reduced in size slightly to fit the book's page format which allows for page headers and footers.

- The samples use font size of 11 points rather than the 12 points I recommend as ideal. Also, the top/bottom and right/left margins have been adjusted to fit the margins of this book. You may want to readjust them for an actual letter.

The above notes are also included as "Template Notes" on the "About the Templates" page for each set of real-life templates that follow in this guide.

REAL-LIFE TEMPLATES FOR SUCCESS

Before getting into the specifics of how to write recommendation letters, it is important to discuss the presentation of the sample letter templates that are included later in this guide.

I am referring of course to the fully-formatted real-life templates included later in this guide for all types of recommendation and reference letters.

All of the sample templates for the various types of recommendation letters contained in this Kit are presented in a format that I call "fully-formatted real-life templates". In fact, it is these real-life templates which make all of my writing kits unique. They are what my customers have told me they love about my writing kits.

Real-life templates are ACTUAL letters written for "real-life" situations. ALL of the letters in this Kit were written for actual real-life situations.

I am convinced that real-life templates are by far the most useful tools for people when they need to draft any kind of document. These templates are a quantum leap beyond the traditional one or two line "fill-in-the-blank" cut-and-paste templates.

The remainder of this section is an expanded and updated version of an article I wrote titled *"Use Real-Life Templates For Writing Success"*. The original version of that article has been published in various eZines and websites across the Net.

REAL-LIFE TEMPLATES FOR BETTER LETTERS

At some point along the way, most of us have used what are commonly called "fill-in-the-blank" writing templates. We might have used them to write a letter, format an essay, or set up a resume or c.v.

What I'm talking about here are those form letter templates that you see in text books and work books, with blank lines and spaces where you're supposed to fill in the appropriate words.

Fill-In-The-Blank Template - Sample

For example, in the case of a letter, a typical "fill-in-the-blank" resignation letter template would look something like this:

Dear [NAME OF RECIPIENT]:

This is to advise you that I will be leaving [NAME OF ORGANIZATION] to occupy a position with [NAME OF ORGANIZATION], as a [NAME OF POSITION], effective [DATE OF DEPARTURE].

As you know, I have been looking for an opportunity in the [NAME OF FIELD] for quite some time now. When I saw that [NEW ORGANIZATION] had a position available I immediately applied and was fortunate to be offered the job.

It will not be easy for me to leave [CURRENT ORGANIZATION]. The company and its people have been an important part of my life for the past [X YRS/MOS.]. At the same time, I cannot pass up a career opportunity like this one that offers a future in the field for which I was trained.

I would like to take this opportunity to sincerely thank you for all of your help and support during the years with [CURRENT ORGANIZATION]. I have no doubt that it was the knowledge and experience I gained working for you that helped me obtain the new position.

Please pass on to [BOSS OF RECIPIENT] and the rest of the senior management team both my regrets about leaving and my gratitude, for what [CURRENT ORGANIZATION] has done for me. I want to assure them that I am leaving with the highest regard for this innovative company.

I wish you and all of my friends and colleagues at [ORGANIZATION] the very best in the future.

Sincerely,

[NAME/TITLE OF ORIGINATOR]

Although this "fill-in-the-blank" approach can work, it has a number of drawbacks:

Disadvantages of Fill-In-The-Blank Templates

- Because of their generic nature, they tend to generalize so much that they resemble a computer-generated form letter.

- They don't provide specific information on how a professional would properly fill in the required information [i.e. BLANK FIELDS].

- The content is typically watered-down, using generic terms in order to try and cover every possible situation.

- They don't provide mental stimulation or show how a professional might word the letter in a specific real-life context.

- They are difficult to work with and virtually useless for 98% of real-life situations, since they lack real-life content.

Real-Life Letter Template - Sample

On the other hand, here's what a "real-life template" of that resignation letter would look like for the same situation covered above:

Dear Sharon:

This is to advise you that I will be leaving Allied Industries Inc. to occupy a position with Telecom Systems International (TSI), as a Customer Service Agent, effective June 30, 20xx.

As you know, I have been looking for an opportunity in the customer services field for quite some time now. When I saw that TSI had a position available I immediately applied and was fortunate to be offered the job.

It will not be easy for me to leave Allied Industries. The company and its people have been an important part of my life over the past four years. At the same time, I cannot pass up a career opportunity like this one that offers a future in the field for which I was trained.

I would like to take this opportunity to thank you sincerely for all of your help and support during the years with Allied. I have no doubt that it was the knowledge and experience I gained working for you that helped me obtain the new position.

Please pass on to Jim Dunning and the rest of the senior management team, both my regrets about leaving and my sincere gratitude, for what Allied has done for me. I want to assure them that I am leaving with the highest regard for this innovative company.

I wish you and all of my friends and colleagues at Allied Industries, the very best in the future

Sincerely,

Jessica Amherst
Corporate Support Group

Actually, I was overly generous with the fill-in-the-blanks template example I gave above. My version gave much more information than is normally included in a typical fill-in-the-blank template. They typically consist of two or three generic statements with a bunch of blanks to fill in. Not much help, in my opinion.

To see direct comparison of real-life templates with fill-in-the-blank models go to: http://instantrecommendationletterkit.com/comparetemps.html

ADVANTAGES OF REAL-LIFE TEMPLATES

Clearly, there can be no doubt that the "model" that most of us would rather work with if we had to write a similar letter is definitely the "real-life" template.

That's because you can relate to it. It talks about real-life people in a real-life situation that you can identify with. And, you get to see exactly how a professional writer worded it in a particular context.

Following, are the main advantages of "real-life" templates.

Content With Value

Working with "real-life" templates is easier. You can adapt them to YOUR actual situations because they provide visual and intellectual cues to which you can relate.

Naturally, when you see how a copywriter or consultant has dealt with a "real-life" scenario, in terms of word choice, context, and punctuation, it is much easier to adapt effectively to the real-life situation for which you are writing. In that way, the actual content has value.

Easy To Work With

"Real-life" templates are just as easy to work with as other templates. You simply load them into your word processing program, and edit and adjust them to fit your own specific situation.

Presto! You have a fully formatted real-life letter ready to be printed and sent out in the mail and you have the comfort of knowing that what you are sending has already been used in other "real-life" situations; and it's grammatically correct.

Real-Life Content

With real-life templates, it is much easier to find an adaptable "fit" for the situation for which you are writing. Not only do they give you the final format of a document, their content provides an excellent real-life sample and gives food-for-thought to assist you in the writing process.

Fully-Formatted Final Versions

"Real-life" templates are fully-formatted as final documents so that you can see exactly what they looked like when they were sent out in "real-life" situations. They don't look like some kind of "draft" computer-generated form letter.

Go ahead; browse the sample letter templates later in this guide. (see page 61).

Are you back yet? Okay.

Now I ask you, would you rather work from a "fill-in-the-blanks" generic template or from a fully-formatted "real-life" template?

I have no doubt that the vast majority of readers would choose the latter for the reasons given earlier. In fact, visitors to my websites have already confirmed this.

Reality Check

As already stated, all of the sample templates presented in this Kit are based on real-life situations using real-life content, for all of the reasons described above.

However, names, addresses, phone numbers, etc. that could be used to identify a specific individual have been altered to protect privacy.

WORKING WITH REAL-LIFE TEMPLATES

Let me take just a minute here to make sure you understand EXACTLY what you are getting with the "real-life templates" in this Kit. Here's the story...

If you bought the eBook version of _Instant Recommendation Letter Kit_, you also received a set of real-life word processing templates. _(If you have the paperback version, go to the Special Preface to find out how to get the templates by e-mail)._

With the templates you have a virtual "writing toolkit" packed with almost 80 fully-formatted real-life recommendation letter templates that you can download straight into your word processing program!

That's right; all of the templates included in this book/ebook have also been given to you in downloadable form in another file (MS-Word compatible).

Here's how you use those templates in a typical situation …

1. You have to write a recommendation letter in a hurry.

2. You check the book and find the template that most closely fits your situation.

3. You open that specific template right into your word processing program!

4. You copy, cut and paste a few revisions to transform the template to fit your own specific situation.

PRESTO! You've got your fully-formatted, professional-quality final letter all ready-to-go in finished form, just like a real professional would write it.

Just think of the writing power this gives you!

No more tedious retyping/recopying from scratch; it's an instant document creator.

That's the power and the beauty of "real-life downloadable templates."

Using real-life templates, it shouldn't take you more than a few minutes to draft professional quality recommendation letters that cover your specific situation.

The section of this guide titled "The Template Adaptation Method" (See page 41) explains, in step-by-step detail, how you can quickly and easily adapt any real-life letter template in this Kit to help you draft a recommendation letter to fit your own specific situation.

<u>WARNING</u>: BEWARE OF AUTOMATIC LETTER GENERATORS

If you spend some time online looking for letter writing help of any kind you will no doubt come across software programs that "automatically generate" various letters.

All I can say about these products is "buyer beware"! Although it might have seemed like a good idea at the time, these "software letter generators" are only slightly better than the typical fill-in-the-bank templates.

Why, you ask? **Here are the problems** with the typical letter-writing software:

- For each letter, you get to choose from a half-dozen or so, **completely out of context**, independent/unconnected pre-written, one-liner phrases that you have to plug into your letter as separate sentences.

- The final software-generated-letter is a **disjointed collection of weakly-linked sentences** and/or paragraphs that actually needs a professional editing job just to make it presentable before sending it out.

- **You do not get "in-context" mental stimulation** and visual cues to help you visualize the final finished product, as you do with a fully-formatted real-life template.

- The final letters come **entirely unformatted** so that you then have to completely set-up and format the letter from scratch before it can be sent.

- Software letter generators **provide little or no guidance** in the way of tips, pointers and information on the best ways to approach writing the different types of letters. *(Trust me, not all letters are equal! There are some specific things you need to know about writing many of the more complex letters).*

To see an actual comparison of real-life templates with two of the top automatic letter generators, check out the following link:
http://www.instantletterwritingkit.com/comparetemps.html

As I said, software letter generators might seem like a good idea in theory, but they just don't do the job of a real-life template.

THE TEMPLATE ADAPTATION METHOD

Just before getting into the actual recommendation templates I'm going to show you how you can easily adapt any template to quickly and effectively create your own recommendation letter, usually in a matter of minutes.

It's a method that I developed for myself years ago, and I now use it for just about everything I have to write. I call it the **Template Adaptation Method**^{TM}.

Please take some time to read and understand the next few pages.

Once you know how to use this method you will be able to apply it to just about anything you will ever have to write, including recommendation letters of course.

USING THE TEMPLATE ADAPTATION METHOD

The Template Adaptation Method^{TM} **is an approach that allows you to use pre-written templates as a tool for developing your own letters, in no time flat.**

Using this method in conjunction with real-life templates you will never have to start from a blank page or screen again. Not only will it significantly reduce the time taken to write your letter, it will also result in a better quality final product.

I'm about to give up a major "trade secret" here.

I first discovered the Template Adaptation Method^{TM} years ago when I was in a job where I had to write a lot of letters and reports. At times, I suffered serious bouts of "writer's block". Sometimes I would stare at a blank page or screen, and/or off into space for days. Yes, literally for days sometimes.

Then one fateful day I had the proverbial "eureka experience" that changed my writing life forever! Really. I'm not sure whether it was by chance, or fluke, or divine intervention, but that particular day I stumbled onto a powerful secret for overcoming "writer's block" -- instantly.

As is usually the case when one makes one of these little "breakthroughs" in life, it was incredibly simple. It was so obvious.

So, here is the big secret that I discovered for beating writer's block:

Place an actual sample of the type of document you have to write, directly in your line-of-sight. The closer the sample is to what you have to write, the better. For example, if you have to write a reference letter, try to find a previous letter of reference that you have written. If you haven't done one before, get one that somebody else has written and post that up.

That's it! It really is that simple. I call it the Template Adaptation Method[TM]**.**

Don't ask me exactly how/why it works, but it does. My theory is that it gives your brain a concrete visual cue as to what you need to write in a very specific way. Staring at a blank page/screen, or out into white space, just doesn't give you this kind of cue.

In fact, the very first thing I did when I sat in front of my computer to write this ebook was to open a copy of one of my previous ebooks right into my word processor. I then placed a hard copy of that ebook on the book stand/easel to my left, right next to the computer monitor.

I was able to immediately start writing the new ebook on the spot using that "real-life template document" in my word processing program, and by referring to the hard copy document beside me, as required. With direct visual access to these two versions of that document, I was never at a loss for words or ideas.

Although the final versions have completely different content, it doesn't matter, because the structure and the flow are similar, which is the key.

Using my Template Adaptation Method[TM] for this ebook was a great time-saver. It took about half the time it would have, if working from a blank page/screen.

Not only that; it also allowed me to get started almost instantly!

So, if you are ever blocked in writing your own letters, staring into space and/or at a blank page/screen, I urge you to find a suitable real-life template and give my Template Adaptation Method™ a try. I guarantee, you'll be pleasantly surprised.

As I stated above, I personally use the Template Adaptation Method™ all the time when I'm writing letters, essays, reports, and just about any other document.

In fact, thanks to this method, it is rare that I ever have to write anything anymore from a blank page or screen. And that's a great relief, I can tell you!

A Step-by-Step Approach

Here's a step-by-step way to apply the Template Adaptation Method™ when developing your own recommendation letters.

1. **First, scan through all of the sample letters** included in this guide and find one that is <u>closest</u> to what you need. Look at ALL the samples just to be sure. For example, if you need to write a recommendation letter you will want to check the reference letter samples as well, to see if there's a better "fit" there.

2. **Once you've found a letter template** "along the lines" of what you're looking for in terms of approach and style, copy and paste it into your word processor.

3. **Start drafting your letter on the same page** as the one you just selected and pasted, <u>one paragraph at a time</u>. This point is very important. Make sure you do it one paragraph at a time.

4. **Proceed through the entire document**, writing your letter, line-by-line, paragraph-by-paragraph, until you have progressed through the entire letter.

5. **Read through the paragraphs you have created** and make sure they make sense and flow smoothly and logically, in step with the flow of the template.

6. **Now, delete all of the template paragraphs.** What will remain is your own personalized letter, but using the approach and style of the template.

On the following pages I'll give you an example that shows exactly how this template adaptation method is applied, in practical terms.

A Real-Life Example

I'm going to give you a detailed real-life example here so you will understand exactly what I'm talking about.

For this exercise I'll use the college application recommendation letter template from page 88 since it's a typical recommendation letter situation.

Here's that entire real-life recommendation letter template:

Dear Ms. Shepperd:

I am very pleased to write this recommendation on behalf of Layla Bell.

Layla has been a student in the accelerated liberal arts program at Holymount High for her entire five years of high school. During that period I have observed her grow into a poised and accomplished young woman. She is an exceptional student with excellent grades resulting from diligent work habits.

Layla has superior interpersonal skills and works equally well independently or in a group setting. She also displays good leadership skills when involved in group projects. She is very well liked and respected by both her peers and her teachers.

Among her many service activities at the school, Layla was a coach of the junior track team for the past two years and she was a member of the senior cross-country team. She also took part in the Mentoring Program and helped a number of juniors navigate their way through their first year of high school. In addition, Layla was involved in organizing a number of fund-raising projects at the school, including a team marathon event that raised over $5,000 for cancer research.

Layla has shown an ongoing interest in world affairs and international development. It is my understanding that she intends to pursue an Honors degree in Political Science or Sociology. She has traveled extensively and has written outstanding reports with observations on conditions she has witnessed throughout the world.

I believe that Layla Bell has tremendous potential as a student and I feel quite confident that she would be an asset to both student life and academics at MacDonald University.

Yours truly,

Allan S. Fenton
Vice-Principal

The above example is a fairly typical recommendation letter written for an undergraduate college admission situation. We are going to use it on the following pages as our "input template" when we apply the Template Adaptation Method™.

Now, here's the scenario for which you need a <u>new</u> recommendation letter:

- You're a high school guidance counselor and you've been asked by one of this year's graduating students to write a recommendation letter that they can use for his application to liberal arts programs at a number of colleges.

- You are quite familiar with this particular student and have nothing but good things to say about them.

- You worked on a number of fund-raising campaigns with this student and want to reflect that in your letter somehow.

Now let's use the above template with the Template Adaptation Method™ to create your new letter, paragraph by paragraph.

Paragraph 1 – _Template Version:_

Dear Ms. Shepperd:
I am very pleased to write this recommendation on behalf of Layla Bell.

Paragraph 1 – New Version:

Dear Admissions Committee Members:
I have been asked by Jeffrey Flowers to write this letter of recommendation in support of his application to your liberal arts program.

Paragraph 2 – _Template Version:_

Layla has been a student in the accelerated liberal arts program at Holymount High for her entire five years of high school. During that period I have observed her grow into a poised and accomplished young woman. She is an exceptional student with excellent grades resulting from diligent work habits.

Paragraph 2 – New Version:

I have known Jeffrey for his entire career at Westview High. Over those four years I have seen Jeff develop from a timid junior into a confident and active senior who has made a significant impact on both the school and the entire community. He has matured into an outstanding and exemplary individual.

Paragraph 3 – *Template Version:*

Layla has superior interpersonal skills and works equally well independently or in a group setting. She also displays good leadership skills when involved in group projects. She is very well liked and respected by both her peers and her teachers.

Paragraph 3 – New Version:

Jeffrey is a natural born leader who has the ability to motivate and involve even the most cynical and apathetic students. He has the rare ability to be able to relate to people at the human level and thus easily gains their confidence and respect. As a result, he was very popular among both students and the teaching faculty.

Paragraph 4 – *Template Version:*

Among her many service activities at the school, Layla was a coach of the junior track team for the past two years and she was a member of the senior cross-country team. She also took part in the Mentoring Program and helped a number of juniors navigate their way through their first year of high school. In addition, Layla was involved in organizing a number of fund-raising projects at the school, including a team marathon event that raised over $5,000 for cancer research.

Paragraph 4 – New Version:

Jeff's many activities at school in his final year included track and field (State finals), school football team (quarterback), and President of the Computer Club. Nevertheless, I was particularly impressed with his community work in which he was the leading student representative in a community-wide campaign that raised over $100K for the new Children's hospital. He worked tirelessly on that project.

Paragraph 5 – *Template Version:*

Layla has shown an ongoing interest in world affairs and international development. I understand that she intends to pursue an Honors degree in Political Science or Sociology. She has traveled extensively and has written outstanding reports with observations on conditions she has witnessed throughout the world.

Paragraph 5 – New Version:

A frequent contributor to the school newspaper, Jeff's articles were always well-written and thoughtful, making it no surprise that he plans to someday pursue a career in journalism. I have no doubt that he will succeed in that or any other career he chooses.

Closing – *Template Version:*

I believe that Layla Bell has tremendous potential as a student and I feel quite confident that she would be an asset to both student life and academics at MacDonald University.

Closing – New Version:

In closing, I have nothing but positive things to say about Jeffrey Flowers and I believe he will be a valuable addition to your school, both in the classroom and in all aspects of college life.

Now let's copy and paste all of the "New Version" paragraphs developed above into a new template and see what it looks like:

Dear Admissions Committee Members:

I have been asked by Jeffrey Flowers to write this letter of recommendation in support of his application to your liberal arts program.

I have known Jeffrey for his entire career at Westview High. Over those four years I have seen Jeff develop from a timid junior into a confident and active senior who has made a significant impact on both the school and the entire community. He has matured into an outstanding and exemplary individual.

Jeffrey is a natural born leader who has the ability to motivate and involve even the most cynical and apathetic students. He has the rare ability to be able to relate to people at the human level and thus easily gains their confidence and respect. As a result, he was very popular among both students and the teaching faculty.

Jeff's many activities at school in his final year included track and field (State finals), school football team (quarterback), and President of the Computer Club. Nevertheless, I was particularly impressed with his community work in which he was the leading student representative in a community-wide campaign that raised over $100,000 for the new Children's hospital. He worked tirelessly on that project.

A frequent contributor to the school newspaper, Jeff's articles were always well-written and thoughtful, making it no surprise that he plans to someday pursue a career in journalism. I have no doubt that he will succeeed in that or any other career he chooses.

In closing, I have nothing but positive things to say about Jeffrey Flowers and I believe he will be a valuable addition to your school, both in the classroom and in all aspects of college life.

Yours sincerely,

Robert Ralston
Career Counselor, Westview High School

Voila! We have a brand new recommendation letter based on the original template. But it's also different, dealing with a totally different recommendation situation.

Clearly, we have two different letters above. It's completely different phrasing; and also completely different specific content. The only things similar between the two letters are the approach/angle and the logical flow.

So, as you can see from the foregoing example, using the real-life template it was an easy matter to quickly adapt the approach, style and contents of the original recommendation letter to create a new one to fit the new situation.

Once you become familiar with using this method you'll find that your adapted letter will start to develop a life of its own. Soon, you'll begin adding things and you'll start plugging in your own words and phrases.

In the end you'll have a very different letter - one adapted to the specific situation you're writing the letter for.

In most cases, your new letter will actually be BETTER than the template you're working from.

The important point being that; by using the Template Adaptation Method[TM] **you don't have to start from a blank piece of paper or computer screen.**

Bottom Line

I cannot over-emphasize how powerful this tempalte adaptation can be, and I urge you to give it a try, even if you're a little skeptical. I can't tell you how many times using this simple technique has "saved my bacon" over the years.

Sometimes, the simplest things are the most effective. This is a case where that definitely holds true.

And the beauty of it is that it can be used for virtually ANY document, from a one-page letter to a 200 page "how to" book. **It's definitely worth a try.**

Here's a summary of the Template Adaptation Method™ we just applied:

Template Adaptation Method™

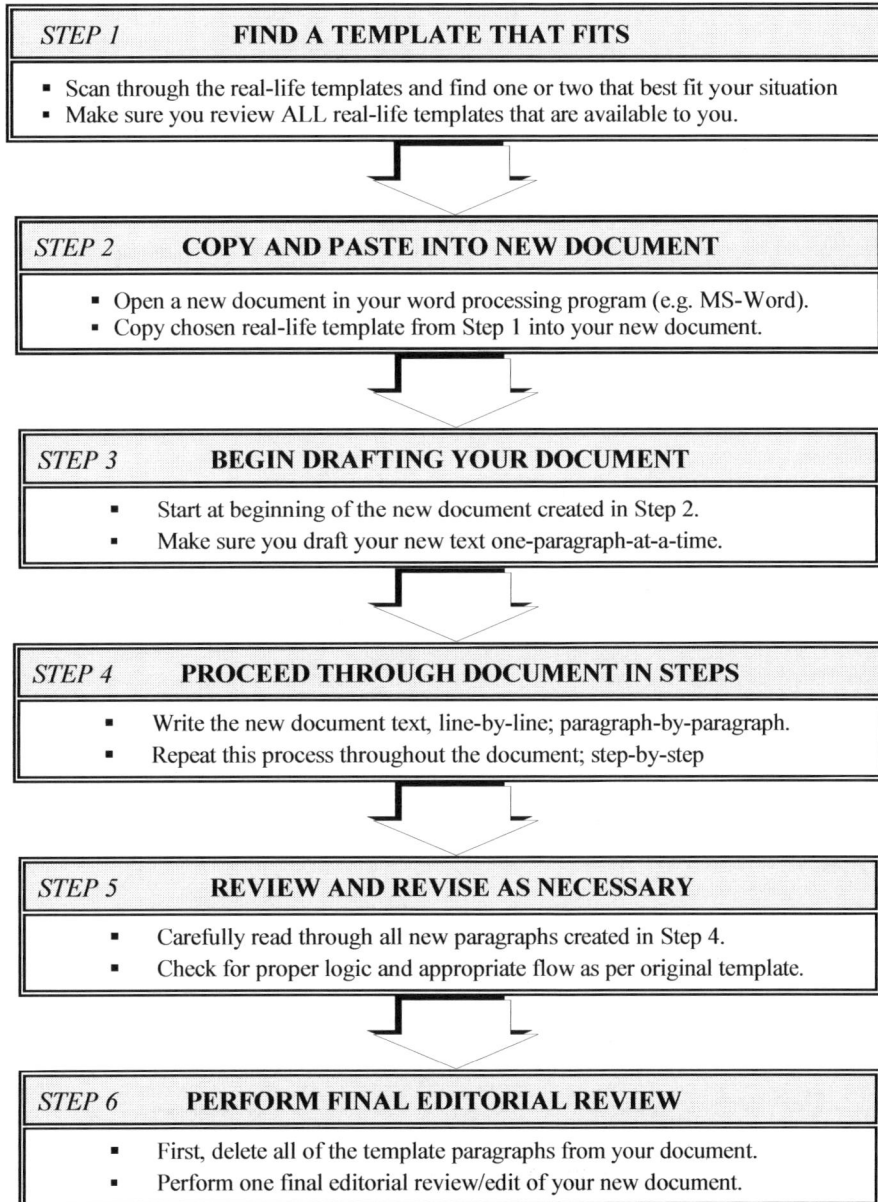

STEP 1	FIND A TEMPLATE THAT FITS
• Scan through the real-life templates and find one or two that best fit your situation • Make sure you review ALL real-life templates that are available to you.	

STEP 2	COPY AND PASTE INTO NEW DOCUMENT
• Open a new document in your word processing program (e.g. MS-Word). • Copy chosen real-life template from Step 1 into your new document.	

STEP 3	BEGIN DRAFTING YOUR DOCUMENT
• Start at beginning of the new document created in Step 2. • Make sure you draft your new text one-paragraph-at-a-time.	

STEP 4	PROCEED THROUGH DOCUMENT IN STEPS
• Write the new document text, line-by-line; paragraph-by-paragraph. • Repeat this process throughout the document; step-by-step	

STEP 5	REVIEW AND REVISE AS NECESSARY
• Carefully read through all new paragraphs created in Step 4. • Check for proper logic and appropriate flow as per original template.	

STEP 6	PERFORM FINAL EDITORIAL REVIEW
• First, delete all of the template paragraphs from your document. • Perform one final editorial review/edit of your new document.	

In the end you'll have a very different letter - one adapted to your own real-life situation. As I said before, in most cases your new letter will be better than the template you're working from.

BENEFITS OF TEMPLATE ADAPTATION METHOD

The main benefits of the Template Adaptation Method™ are the following:

- Downloading a template into your word processor and developing a new letter, paragraph by paragraph, saves significant time over starting from scratch.

- Using a pre-written real-life template simplifies the process of finding an approach for your letter.

- The template will stimulate your thinking process and will give you new ideas for your letter.

- The template will act as a "checklist" to make sure you've covered everything that you need to.

- You know that you're using a "model" that has been used successfully by others.

In short – the Template Adaptation Method™ will give you a superior letter.

So, if you ever struggle with getting started writing your letters, I highly recommend that you try this proven method. It really will simplify your letter-writing process.

The important point being that; by using the Template Adaptation Method™ you will never again have to start writing a new document from a blank piece of paper or computer screen.

RECOMMENDATION POWER PHRASES

If you closely review the recommendation and reference letter templates included in this Kit you will notice that there are certain "types" of phrases that recur over and over again throughout the various letters.

These statement types can be grouped into three major categories: opening statements, assessment statements, and closing statements.

Opening, assessment, and closing statements are the three key components to any recommendation or reference letter. If you formulate and link these well, you are guaranteed to have an excellent letter.

To help you quickly formulate your own recommendation letters I have compiled a list of more than 150 generic "Recommendation Letter Power Phrases" that you can use. These statements are grouped into the three categories I just listed above.

OPENING STATEMENTS

The opening statement in a recommendation letter should state the name of the person being recommended. It can also explain why you are the person writing the letter. The opening statement should normally be one short sentence and should never exceed two sentences.

Following are more than 50 generic typical opening statements for inclusion in recommendation letters and reference letters:

- I am writing this letter at the request of [name of requestor].

- This is in reply to your recent letter to me in which you asked my opinion of [name].

- [name of requestor] requested that I write this letter to you.

- This is in response to your recent request for a letter of recommendation for [name].

- I am pleased to be able to write this letter of recommendation for [name].

- I was asked by [name of requestor] to write this letter to you.

- I had the fortunate opportunity to work closely with [name] for x years…

- [name of requestor] asked me to write this to you based on our working association…

- [name] worked under my direct supervision for…

- Thank you for your letter asking me for my assessment of [name].

- I am very pleased to write this recommendation on behalf of [name].

- It is an honor for me to be able to write this letter of recommendation on behalf of my former student [name].

- This letter is in support of [name of requestor]'s application for admission to…

- I am writing this letter in support of [name of requestor]'s nomination for…

- This is to confirm that [name] Rocco worked under my direct supervision from…

- The purpose of this is to provide a character reference for [name of requestor].

- I have been asked by [name of requestor] to write this letter of reference because our company will no longer be operating…

- The purpose of this is to explain why [name of requestor] left our company after less than one year of employment.

- The purpose of this letter is to officially commend [name] for his exceptional contribution to…

- I am providing this reference letter at the request of [name of requestor], who has asked me to serve as a reference on her behalf.

- I am writing this letter of reference at the request of [name of requestor].

- I have been asked to provide this letter of reference by [name of requestor].

- The purpose of this letter is to inform you of one of our citizens, [name of nominee], who I believe should receive special recognition for…

- I am writing this letter of reference at the request of [name of requestor] who is applying for your graduate program in [discipline].

- I am writing this reference letter in support of [name of requestor]'s application for one of your entrance scholarships.

- The purpose of this is to provide a character reference for [name of requestor].

- I am pleased to respond to [name of requestor]'s request for a letter of reference about [another name].

- I have been asked by [name of requestor] to write this reference letter.

- I am writing this letter of reference to state how grateful my family is that [name] was assigned to us.

- I am writing this in support of [name of nominee]'s nomination for...

- This letter is in support of [name of applicant]'s application for admission to your graduate program in [discipline].

- I am very pleased to write this recommendation letter in support of [name of applicant]'s application for admission to...

- I am writing this letter for [name of applicant] as one of the requirements for his admission to...

- It is a pleasure for me to be able to write a letter of recommendation supporting [name of applicant]'s application for...

- I have been asked to write this letter for [name of applicant] as one of the requirements for her admission to...

- The purpose of this is to thank you for the positive influence you have had on our daughter [name of student] this past school year.

- The purpose of this is to submit the name of [name of nominee] for the...

- I am providing this reference letter at the request of [name], who has asked me to serve as a reference on her behalf.

- Please be advised that the information contained in this letter is confidential and should be treated as such.

- I am writing this letter of reference at the request of [name] who I understand has applied for the position of [position name] with your company.

- I am writing this letter of recommendation to convey to you my professional and personal assessment of [name] as an employee.

- It is with mixed feelings that I write this confidential recommendation to you about [name].

- [name] was a member of my sales team for a little over two years between June 20xx and September 20xx.

- I got to know [name] over an 18 month period while he worked for me as a consultant on a project sponsored by [organization].

- I am writing this letter of reference at the request of [name] who is applying for your graduate program in [name].

- [name] demonstrated his honesty and integrity just recently when he…

- I am writing this reference letter in support of [name]'s application for one of your entrance scholarships for first-year undergraduate students.

- The purpose of this is to provide a character reference for [name] who I have known as a client and friend for xx years.

- I have known and worked closely with [name] since 19xx. During that time, I have developed the highest esteem for her, both personally and professionally, and it is therefore my pleasure to write this letter on her behalf.

- I have been asked by [name] to write this reference letter on his behalf because he has been forced to move out of State for family reasons.

- [name] was a student teacher under my supervision for xx weeks, from January xx, 20xx to April xx, 20xx at [name of school].

- It is with great pleasure that I recommend [name] for acceptance into your MBA program.

- It is a pleasure for me to be able to write a letter of recommendation supporting [name]'s application for admission to your Ph.D. program in [subject name].

ASSESSMENT STATEMENTS

Assessment statements in a recommendation letter are those sentences and phrases that contain the recommender's specific assessment of the performance, characteristics, and attributes of the person being recommended.

Normally an assessment statement will be followed up by one or two specific performance-related examples backing up the statement just made.

Following are more than 50 generic typical assessment statements for inclusion in recommendation letters and reference letters:

- [name] is a hard-working self-starter who invariably understands exactly what a project is all about.

- [name] is a resourceful, creative, and solution-oriented person

- [name] has superior written and verbal communication skills.

- [name] consistently produces high quality work in a timely fashion.

- The only area of weakness that I ever noted in [name]'s performance was…

- [name] consistently produced superior quality work, on time, and within budget targets.

- Overall, [name] is a conscientious, dedicated, and exemplary employee.

- As both a co-worker and colleague, [name] was outstanding.

- [name] performed extremely well in what was often a multi-tasking situation.

- On the personal attribute side, [name] was truly a pleasure to work with.

- I believe that [name] consistently demonstrated that she has the potential to perform even more complex and demanding duties.

- [name] was co-operative and positive at all times and was very well-liked by both her colleagues and the senior managers.

- One particular strength that [name] demonstrated frequently, was her leadership ability.

- [name] is a cooperative and willing employee, always ready to go the extra mile.

- [name] has an amazing ability to relate to people from all walks of life.

- I'm sure that [name] would be a valuable addition to your committee.

- [name] has superior interpersonal skills and works equally well, independently, or in a group setting.

- In my opinion [name] is an exceptionally bright and hardworking individual who throws herself enthusiastically into whatever she undertakes.

- I believe it is [name]'s strong work ethic, coupled with her meticulous attention to detail that allows her to produce such high quality papers.

- [name]'s ability to work well with others, even under stress, allows her to get the most out of group assignments.

- As an individual, [name] is an extremely well-rounded person.

- Based on the foregoing it should be clear that I regard [name] as an outstanding individual in every sense of the word.

- One of [name]'s key strengths is his ability to analyze, integrate, and synthesize information.

- I believe [name] has excellent potential to become a high-performing sales professional.

- I really can't think of anything of consequence on the negative side of the personality ledger when it comes to [name].

- Based on my knowledge of her, I would not hesitate to state that [name] is a highly dependable individual of seemingly impeccable character.

- As a student, [name] was a hard-working and highly committed to his education.

- As a friend, [name] is a standout. He is a loyal, honest, considerate, and supportive individual.

- [name]'s contribution has been outstanding, and exemplifies the qualities of excellence, and professionalism that are embodied in the [title of award].

- In light of the foregoing assessment of your academic and related activities as a faculty member, I find your overall performance over the past academic year to be unsatisfactory and in need of significant improvement.

- Overall, based on her academic performance in my courses, and my understanding of the [position name] position for which she is applying, I believe that [name] would perform an adequate job.

- I would say that [name] is very well-suited to the job.

- Such a position would seem to be the next logical step for [name] in his career development in the [field/discipline] field.

- In summary, based on our time working together, I would rate [name] as an intelligent, well-organized and hard working professional who brings a lot of enthusiasm to his work.

- I believe that [name] would be a welcome addition to any graduate program in [discipline].

- I believe that [name] would be an excellent candidate for any one of the scholarships you are offering.

- I would ask that you please consider awarding one of your scholarships to [name] in order to supplement her education.

- During that time, I have developed the highest esteem for [name], both personally and professionally, and it is therefore my pleasure to write this letter on her behalf.

- [name] is an extraordinary thinker, instructor, and friend. She has my gratitude and my respect, and I recommend her services wholeheartedly.

- In short, I believe that [name] has the talent, willingness and ability to be a highly productive staff member in the area of ...

- [name] is a high energy self-starter who quickly assumes responsibility and is not afraid to face new challenges and situations.

- [name] also demonstrated strong leadership abilities in his group project assignments.

- Based on the foregoing, you can see that I regard [name] as an outstanding student and individual in every sense of the word.

- [name] is a personable, teachable, hard working and highly intelligent student.

- I believe that [name] is an exceptionally bright and hardworking individual who throws himself enthusiastically into whatever he undertakes.

- [name] is a resourceful, creative and solution-oriented person who is frequently able to come up with new and innovative approaches to his assigned projects.

- [name] also demonstrated above-average administrative capabilities, consistently submitting his sales and trip reports on time, and completed accurately.

- I would rate [name] as an intelligent, well-organized and hard working professional who brings a lot of enthusiasm to his work.

- [name] regularly demonstrates empathy and compassion for those less fortunate.

- I cannot speak highly enough of [name]'s skills and capacities. In addition to being a brilliant analyst, she is a wonderful teacher - attentive, patient, and thoughtful.

- Over the past year and one-half [name] has demonstrated to me, time and again, that he is an exceptional communicator and a gifted teacher.

- [name] performed exceptionally well in each class he took with me. He is an intelligent, enthusiastic, well-prepared, well-spoken, and assertive individual who gets a lot out of his classes and also gives back a considerable amount through his active participation.

- As an individual, [name] appears to be an extremely well-rounded person. He is likeable, straightforward, down-to-earth, confident, and generally, a pleasant person to be around.

- Throughout her years with us, [name] has consistently displayed a remarkable ability to handle frustrating situations with patience, tact, and diplomacy.

- Two hallmarks of [name]'s character are her honesty and integrity.

- The only area of weakness that I have ever noted about [name] is her lack of formal training.

- I believe that [name] possesses a rare blend of superior analytical, technical, and interpersonal skills.

CLOSING STATEMENTS

The closing statement in a recommendation letter should be one or two sentences at most and it should make a clear statement of recommendation that flows logically from the points made in the assessment part of the letter. They will often begin with transition phrases such as: "In summary,…", "In Closing,…", "Based on the foregoing,…", "Accordingly,…", etc.

Closing statements in recommendation letters are generally positive, but in some circumstances they may be qualified, or even completely negative.

Following are more than 40 generic typical closing statements for inclusion in recommendation letters and reference letters:

- I can unreservedly recommend [name] to you for…

- I have no hesitation in recommending [name] for…

- Although I respect [name] as a professional, I must say in all honesty that I cannot recommend him for your committee.

- I am very pleased to be able to recommend [name] for…

- Based on my time working with [name], I recommend her very highly for…

- I am pleased to recommend [name] for whatever position she may qualify for at your company.

- I would highly recommend [name] for positions similar to those she held with our organization.

- Without hesitation I am pleased to recommend [name] for…

- I am pleased to be able to recommend [name] as a member of your new committee.

- I believe that [name] has tremendous potential as a student and I feel quite confident that she would be an asset to both student life and academics at…

- I have no doubts about [name]'s abilities and wholeheartedly recommend her…

- I have no hesitation whatsoever in recommending [name] as a participant in…

- I am pleased to recommend [name] to you without reservation.

- Without hesitation I recommend that [name] be chosen as the recipient of…

- In my opinion, [name] has unlimited potential and will eventually go far in…

- I am pleased to recommend your appointment for the coming academic year.

- I am pleased to recommend that you be reappointed as [title of position] for the upcoming academic year.

- Based on my eight months working with him, I am pleased to recommend [name] for any position for which he may qualify.

- I strongly recommend [name] for any position for which technical expertise, analytical skills, speed and accuracy, are primary considerations.

- I trust that you will give serious consideration to selecting [name] for the position of [title of position].

- I believe that [name] would make an excellent addition to your organization.

- Based on my impressions of [name] during the two years he reported to me, I would not hesitate to recommend him as a valuable addition to any sales team.

- I highly recommend [name] for whichever position he applies.

- I believe that [name] will be a valuable employee for whichever company hires her to perform functions in her areas of expertise.

- I would very definitely rehire [name] given the opportunity, and I can therefore recommend him for your consideration.

- I believe that [name] will make an excellent teacher if she decides that is the career she would like to pursue.

- I am sure that with his level of commitment, coupled with continued experience and training, [name] will be a fine addition to…

- I highly recommend the services of [name] should you need a professional counselor/advisor to guide you through a family crisis.

- In closing, I am pleased to recommend [name] without reservation for…

- It is with sincere conviction that I enthusiastically recommend that [name of nominee] be selected as a recipient of [name of award].

- Accordingly, I highly recommend that you accept [name] into your graduate business program.

- It is with great pleasure that I recommend [name] for acceptance into your [program name] program.

- I would greatly appreciate it if you would favorably consider [name]'s application to the [program name] program.

- Accordingly, I can enthusiastically recommend [name] as a prime candidate for your business school.

- I believe that [name] has excellent potential to become a high-calibre senior manager in the future and a credit to the school which he attends.

- I have no hesitation whatsoever in recommending [name] as a participant in your Masters Program on behalf of our company.

- I believe that, given his demonstrated skills, abilities and experience, [name]'s participation will prove beneficial for both your company and the university.

- Because [name] is one of those rare students who clearly has the potential to make a difference, I cannot recommend her more highly.

- I therefore recommend [name] to you for future assignments in his field.

- Based on the foregoing, I cannot in good conscience recommend [name] to you as an employee.

As stated earlier, the foregoing statements are provided here to help you quickly formulate your own recommendation letters if you can't find a template in the Kit that is appropriate for your situation.

Nevertheless, I urge you to make sure you look through ALL of the templates before you try to write a recommendation letter from scratch. There are very few situations to which the templates in this guide can't be easily adapted.

Please refer to the section that explains the Template Adaptation Method in the following section for step-by-step instructions on how this is done.

RECOMMENDATION LETTER TEMPLATES

As mentioned more than once previously, there is often a lot of confusion as to what constitutes a real recommendation letter.

Recommendation letters are often referred to in a number of different ways including: letters of recommendation, reference letters, letters of reference, commendation letters, and sometimes even, performance evaluation letters.

This terminology can be quite confusing, especially when these terms are often used interchangeably, sometimes to mean the same thing, sometimes to mean something different.

TYPES OF RECOMMENDATION LETTERS

There are two primary categories or types of recommendation letters as follows:

- Employment-Related
- College Admission-Related

Employment-Related Recommendation Letters

An employment-related recommendation letter is one that is specifically requested by the person the letter is being written about. It is specifically written by the recommender to assist the person being recommended in their application process for a particular job.

College Admission-Related Recommendation Letters

College admission related recommendation letters, or letters of recommendaton, are written by a requested recommender to support the application of the person being recommended in support of their application to a college or university program.

So, to be absolutely clear; in this section we are referring specifically to the recommendation letters (i.e. letters of recommendation) that are either employment-related or college-related.

WRITING RECOMMENDATION LETTERS

The information in this section is very similar to the writing tips listed later under employment- and college-related reference letters. Nevertheless, there are some subtle differences specific to recommendation letters. So, review them carefully.

As discussed earlier, I choose to repeat some of the information here that was covered earlier so that this could become a stand-alone section and you won't have to flip back-and-forth and get even more confused between reference letters and recommendation letters. (It's already confusing enough, isn't it!)

The following pages define and summarize the characteristics of these types of recommendation letters and provide sample templates for various real-life situations.

EMPLOYMENT-RELATED RECOMMENDATION LETTERS

An employment-related recommendation letter is one that is specifically requested by the person the letter is being written about. Such a letter is normally positive in nature, and written by someone who knows the subject well enough to comment on the skills, abilities, and specific work attributes of that person.

Typically, an employment-related recommendation letter conveys one person's view of the work performance and general workplace demeanor of a person who has worked under their direct supervision. The requestor of the letter normally requires it when applying for a promotion or a new job.

These letters are usually addressed to a specific person to whom the requestor has been asked to submit the letter.

Drafting Tips – Employment-Related Letters

If you are asked by someone to write a letter of recommendation for them, here are a few important points to keep in mind:

Make Sure You Are Comfortable

If you feel that you don't know the person well enough you should decline, stating that as your reason. Also, you may find that it would be difficult for you to say very much positive about the person requesting the letter. However, if you are the requestor's direct supervisor (or former supervisor) you really don't have a choice, but you will have to be honest. Fair, too!

Define Background Parameters First

Before starting to write, make sure you know exactly to whom you are writing. Knowing the name and specific position of the addressee will help the mental process while drafting the letter. In the opening paragraph provide all of the background information such as: relationship to the person, organization, position titles, and time-frame and dates covered by your assessment, and other relevant overall background information.

Ask For Input From the Requestor

If you feel you need more information, don't hesitate to ask the requestor for a copy of their resume/cv. In addition, if you have access to them, you may want to have a quick review of recent performance evaluations. Finally, you can ask the requestor to jot down some key points they would like you to mention (at your discretion). This would include such things as highlighting their work and accomplishments on a special project or task force, etc.

Make Sure You Cover the Entire Person

Your letter should address all of the areas of the person that are job performance related. This would include such things as: basic job performance, motivation level, communication abilities, flexibility, adaptability, energy level, quality of outputs, initiative, leadership, goals achievement success, teamwork, etc. Always try to give a specific example rather than just making an open-ended generalization.

For example, "Shirley's exemplary performance as a Task Force Team Leader in the Restructuring Project highlighted her superior leadership and communication abilities" rather than, "Shirley has superior leadership and communication skills."

Avoid Controversial Statements/Terminology

Do not make statements that you cannot clearly support with facts. For example, statements such as "I believe Frank displays this tendency because he is the product of a dysfunctional background" are clearly not acceptable.

Make sure you avoid using any words or terminology that could be deemed discriminatory such as: race, color, religion, political affiliation, sex, sexual orientation, age, physical appearance, handicaps, marital status, etc.

Use Active, Powerful Words

Neutral words such as good, nice, satisfactory, fair, reasonable, etc. should be avoided. Use active, descriptive words and terms such as: intelligent, assertive, initiator, self-starter, motivated, hard-working, multi-tasker, cooperative, productive, creative, articulate, leader, communicator, team player, innovative, effective, efficient, honest, dependable, mature, etc.

Review the Final Product

Before signing your letter, do a final careful review. Check all spelling and grammar and make sure the terminology used is appropriate. Read it out loud to yourself and imagine being the recipient. Is it fair and balanced? Does it truly convey what you believe and want to say about the person who you are recommending? Make revisions as appropriate.

SAMPLE TEMPLATES - EMPLOYMENT RECOMMENDATION LETTERS

The following pages contain thirteen (13) real-life templates of employment-related recommendation letters for various situations.

Notes Re: Sample Letter Formats

- Although the sample templates in this guide are based on actual situations involving real people, identifying details have been altered to protect privacy.

- All of the sample templates in this guide have been reduced in size slightly to fit the book's page format which allows for page headers and footers.

- The samples use font size of 11 points rather than the 12 points I recommend as ideal. Also, the top/bottom and right/left margins have been adjusted to fit the margins of this book. You may want to readjust them for an actual letter.

Recommendation 1: Employment – Marketing

(print a Recommendation Letter on company letterhead paper)

July 30, 20xx

Mr. Rodney Sims
Director, Marketing Services
Newport Industries Inc.
1500 Elm St.
Dallas, TX 75270

Dear Rodney Sims:

RE: Employment Recommendation – Maria Fuentas

This is in response to your recent request for a letter of recommendation for Maria Fuentas who worked for me up until two years ago.

Maria Fuentas worked under my direct supervision at Jasminder Technologies for a period of five years, ending in May 20xx. During that period, I had the great pleasure of seeing her blossom from a junior marketing trainee at the beginning, into a fully functioning Marketing Program Co-Ordinator in her final two years with the company. This was the last position she held before moving on to a better career opportunity elsewhere.

Maria is a hard-working self-starter who invariably understands exactly what a project is all about from the outset, and how to get it done quickly and effectively. During her two years in the Marketing Co-Ordinator position, I cannot remember an instance in which she missed a major deadline. She often brought projects in below budget and ahead of schedule.

Ms. Fuentas is a resourceful, creative, and solution-oriented person who was frequently able to come up with new and innovative approaches to her assigned projects. She functioned well as a team leader when required, and she worked effectively as a team member under the direction of other team leaders.

On the interpersonal side, Maria has superior written and verbal communication skills. She gets along extremely well with staff under her supervision, as well as colleagues at her own level. She is highly respected, as both a person and a professional, by colleagues, employees, suppliers, and customers alike.

Two years ago, when Ms. Fuentas announced her resignation to take up a new position with a larger company, we were saddened to see her leave, although we wished her the greatest success in her new undertaking. Even now, two years after her departure, I can state that her presence, both as a person and as an exemplary employee, is still missed here.

In closing, as detailed above, based on my experience working with her, I can unreservedly recommend Maria Fuentas to you for any intermediate to senior marketing support position. If you would like further elaboration, feel free to call me at (416) 765-4293.

Sincerely,

Robert Christie
Director, Marketing and Sales

Recommendation 2: Employment – Supervisor

(print a Recommendation Letter on company letterhead paper)

June 28, 20xx

Mr. Stewart Hamilton
Project General Manager
Airport Expansion Project
Grantley Adams Int'l Airport
Christchurch, Barbados, West Indies

Dear Stewart Hamilton:

I am writing this letter at the request of David Handridge who has recently applied for the position of Graphic Design Supervisor on your project.

David worked under my direct supervision in the position of project team Graphic Artist and Designer for a three-year period ending in 20xx. Overall, I would say that he was an excellent employee who fulfilled the duties of his position in an exemplary way.

As an artist and designer, David was first rate, consistently producing high quality work in a timely fashion. When creative work was required, his work was always innovative and fresh. On the technical and engineering side he also produced good accurate work, even though he much preferred the creative assignments. I think it is fair to say that David never let his creative passion get in the way of his technical obligation.

The only area of weakness that I ever noted in David's performance was in his supervisory skills. In his position with the project team he had two junior artists reporting to him. In addition, he was occasionally called upon to supervise contracted graphic artists on larger rush projects.

The problem that came up on a number of occasions was David's inability or reluctance to fully delegate work to his staff. We discussed this issue a number of times and our conclusion was that, David being a perfectionist by nature, felt he had to have a direct hand in all of the work of his subordinates, no matter how routine. Needless to say, on occasion this tendency did cause some problems with staff and had some repercussions on meeting overall project deadlines.

When David approached me about writing this letter, we did discuss this one area of weakness. He now assures me that, in the three years since we worked together, he has completely overcome this one shortcoming through a combination of training and ongoing practical experience.

This being the case, I have no hesitation in recommending David Handridge to you for the position of Graphics Design Supervisor on your team.

I should add that, on the interpersonal side, David is a wonderful person and a good communicator who gets along with people very well, even under heavy deadline pressures.

Should you require further elaboration, don't hesitate to contact me at (416) 742-1594.

Sincerely,

Roger Neilson, CMA
Partner, ADN Engineering Systems Inc.

Recommendation 3: Employment - Physician

(print a Recommendation Letter from a professional on corporate letterhead stationery)

May 28, 20xx

Phalen Emergency Center
John K. Linden Hospital, Stag Hill
Attention: Joshua A. Halton, D.O.
17251 S. 13th Avenue
Phoenix, AZ 85027

Dear Dr. Halton:

<u>**Reference: Christie Sweet, D.O.**</u>

I am writing this letter of reference in support of Christie Sweet, D.O. to work at your center as an emergency medicine physician. I have known Christie in a professional capacity for three years and feel very comfortable in recommending her.

Christie is currently a third year emergency medicine resident and she will be available at the completion of her training in July, 20xx. She has worked under my supervision in a residency program at University of Washington's Shock Trauma Unit. That unit is a high volume, high acuity, Level II, multi-cultural, multi-ethnic, urban community emergency department and includes trauma training. I believe that Christie's high degree of personal commitment to public service coupled with her strong training background provide her with the foundation and skills necessary to handle the challenges of a progressive and rapidly growing emergency department such as the one at John K. Linden, Stag Hill.

Ms. Sweet has asked me to refer her to you at John K. Linden, Stag Hill for a number of specific reasons. First, she had the good fortune to be scheduled for some clinical clerkships at JKLH as a medical student, during which she enjoyed the interactions with staff there and benefited from numerous learning opportunities. Secondly, she had some of the emergency physicians in your group as preceptors during rotations at Kingston Regional Medical Center and she found them to be excellent clinicians and teachers. Lastly, I understand that Christie originates from that area and would like to renew regular contact with her family and friends.

Christie is among the top residents under my supervision, so I can therefore recommend her to you without reservation. I believe she will make an excellent physician.

Please feel free to contact me either by e-mail or phone at any time if you require clarification or have any questions. Contact information is in the letterhead.

Yours sincerely,

Donald Drummond, M.D.

Recommendation 4: Employment - Occupational Therapy

(print Recommendation Letter on company letterhead paper)

April 20, 20xx

Shannon Marchant
Admissions Officer
University of Texas Medical Branch
Department of Occupational Therapy
301 University Blvd.
Galveston, TX 77555

Dear Ms. Marchant:

I have been asked to provide my assessment of Karen Strange in support of her application to your Occupational Therapy program. As requested in your guidelines, I am submitting this to you directly without the intervention of the applicant.

I have had the pleasure of knowing Karen Strange for approximately three years. During that time she has worked for me as both my Personal Assistant and and in parallel with that, as Manager of the Insurance and Shipping departments at Kranston Collectibles Inc. As my assistant, she is required to perform numerous administrative tasks, as well as dealing with customers and supervising the company's rare objects inventory. In her shipping management job, Ms. Strange is responsible for carefully insuring and shipping expensive rare object shipments. In performing these jobs, she has consistently demonstrated that she is a well-organized, efficient, reliable, and responsible individual.

Ms. Strange has demonstrated numerous strengths that I believe will support her in a career as an Occupational Therapist. She is a highly dedicated employee who is able to work under extreme pressure and tight deadlines with calmness and composure. She has the ability to focus on a task and follow it through diligently until it is properly completed. Her level of integrity is beyond reproach. Working in the rare objects business she understands how important it is to respect the privacy of her clients, an attribute that I am sure will also be helpful when dealing with clients in the field of Occupational Therapy.

Karen is always punctual and has excellent computer and verbal communication skills. She is well-respected by both her clients and co-workers and is known for her strong leadership skills. Ms. Strange also has an outstanding ability to deal with different types of people and customers on a one-to-one basis, and she also works well in group situations.

As detailed above, I believe that Karen Strange is an exceptional employee who exceeds expectations in just about every area imaginable. She is highly motivated, constantly striving to get better. For example, the one area where she needed some improvement was in her written communication skills. So on her own, she registered in college and has been taking evening courses to improve her writing skills. In my opinion, Karen's work ethic, coupled with her ongoing efforts towards self-improvement, give her unlimited potential in just about any field she may choose, including Occupational Therapy.

Ms. Strange also has excellent interpersonal skills. On a regular basis she was required to interact with management, co-workers and those reporting to her and always did so with tact and diplomacy. Whether it was collaboration on a project, a presentation, or a rare objects show, in each circumstance, her style of relating to her peers and her interpersonal skills in general were exceptional. Sarah is a kind, warmhearted, and generous individual who has the intuitive ability to deal effectively with a wide variety of people and situations. She also has the capacity to lead and take charge of a group. On the other hand, she has no problem stepping into the background and working effectively from there when it is the best approach for a particular situation. Above all, Karen Strange is highly respected, admired and liked by her peers and is therefore a pleasure to work with.

Accordingly, without reservation, I am pleased to be able to recommend Karen Strange as someone who I believe will be an excellent addition to your Occupational Therapy program. Please feel free to contact me by phone at xxx-xxx-xxxx if you require any further information.

Sincerely,

Kenneth Whyte

Recommendation 5: Employment – Graphic Artist

(print a Recommendation Letter on company letterhead paper)

August 31, 20xx

Ms. Helena Nearington
Executive Editor, Corporate
Griffon Communications Inc.
122 Sixth Avenue,
New York, NY 10011
(212) 632-1789

Dear Ms. Nearington:

Jeremy Cook requested that I write this letter to you with reference to his recent application to your company as a Senior Graphics Illustrator.

Jeremy worked under my supervision as a Graphic Artist from March 20xx, until April 20xx. His responsibilities included conducting research, developing concepts, preparing concept boards, and rendering final graphic illustrations for publication. In addition, he performed some administrative and clerical duties related to the maintenance of the company illustration database.

During the over three years that he reported to me, Jeremy proved himself to be a highly skilled and exceptionally talented graphic illustrator. He consistently produced superior quality work, on time, and within budget targets.

I was particularly impressed by Jeremy's ability to complete all of his work on time, even when he was multi-tasking on two or more major projects in parallel. Frequently, he even completed his assignments ahead of time. His research was invariably thorough and comprehensive, and his editorial concepts were consistently creative and always appropriate and timely.

Jeremy developed his own personal "signature" style of illustration that quickly became the preferred choice of a number of our most valued customers. Because of that, he was assigned as lead illustrator on a number of major projects where normally, he would have been a secondary resource. This gave him the opportunity to demonstrate his excellent team leadership and project management skills.

Overall, Jeremy is a conscientious, dedicated, and exemplary employee. I certainly think he has a bright future ahead of him in the editorial illustration field. Accordingly, I am pleased to recommend him for any intermediate to senior level position as an editorial graphic illustrator.

Sincerely,

Raymond S. Eaton
Director, Graphics Services

Recommendation 6: Employment – Project Support

(print a Recommendation Letter on company letterhead paper)

July 27, 20xx

Ms. Roberta Carnavon
Director, Administrative Services
Corporate Connection Inc.
128 Fifth Avenue
New York, NY, 10011

<u>**RE: Recommendation – Ingrid Heintzman**</u>

Dear Ms. Carnavon:

Ingrid Heintzman worked under my direct supervision for the six-month period, June through November 20xx. Her role was that of Project Coordinator for the planning and implementation of a major international conference that was held in Toronto from October 8-10, 20xx. On that project, I believe that I was able to develop a pretty fair idea of Ingrid's professional capabilities, as well as her primary personal attributes as they relate to the job.

Although relatively inexperienced in project management when she started, Ingrid was very quick to catch on to the required tools and techniques. Using state of the art project management software, she developed an extensive and detailed Project Work Plan which was the basic tool for project planning and status reporting/tracking. In addition, Ingrid developed a Project Accountability Matrix which became the primary day-to-day operational tool for assigning and monitoring staff resources assigned to the various project tasks.

As Project Coordinator, Ingrid was responsible for organizing and chairing daily project review/status meetings. She performed this role with ease and was always well-prepared for the meetings. In addition, Ingrid functioned very well as a member of the project team and consistently dealt with difficult situations appropriately. She is well-organized and is always willing to put in extra time when required.

In addition to the above-mentioned project work, Ingrid was also asked to review and evaluate the Board's Contacts Database. On this project, she did an excellent job of supervising the two employees engaged in updating and modifying this important corporate database. I was particularly impressed with the thorough, well-written, and professional report that Ingrid prepared on her assessment of the database at the end of that project.

As a co-worker and colleague, Ingrid was outstanding. She was co-operative and positive at all times and was very well-liked by both her colleagues and the senior managers she dealt with. She has a very pleasant and positive demeanor and is a pleasure to work with.

In closing, based on my experience working with Ingrid Heintzman, I would recommend her very highly to you. I believe that her presence at the Junior/Intermediate Officer level would greatly benefit your organization. For further information, please contact me at (613) 772-4957.

Sincerely,

Robert Brady, M.B.A
Executive Director, Corporate Services

Recommendation 7: Employment – Co-Worker

(print a Recommendation Letter from a private citizen on regular stationery)

Beverly C. Taylor
3200 75th Avenue
Landover, MD 20785
(301) 731-4754

December 15, 20xx

Ms. Marilyn Joyner
Chief, Personnel Services
Liston Electronics Corp.
185 Parkway Dr.
Landover, MD, 20780

Dear Marilyn Joyner:

RE: Employment Recommendation – Ms. Francis Beresford

I was asked by Francis Beresford to write this letter to you based on my experience working closely with her for most of this year.

I had the fortunate opportunity to work closely with Francis Beresford for eight months earlier this year while she was employed at Brenkman Systems Inc.

Francis occupied the office adjacent to mine, and we collaborated on a number of projects. During the entire period, she was a pleasure to work with. I observed her to be a solid team player, a sincere and hardworking individual, as well as an excellent overall co-worker.

Francis worked very hard at all times, and frequently did more than her share of the work on our joint projects. She was an independent thinker and innovator, and often took the initiative to move things forward while we were waiting for senior management decisions.

I was happy to be able to work with Francis, since management invariably praised the projects that we completed as a team. In fact, on more than one occasion I made specific requests to work with Francis although I had been scheduled for other assignments.

Francis was well-liked by everyone in our office, both co-workers and management. She often went out of her way to be friendly and helpful to her colleagues.

Based on our eight months working together, I am pleased to recommend Francis for whatever position she may qualify for at your company. I believe she is the type of employee that any organization would be happy to have on their workforce.

Sincerely,

Beverly C. Taylor

Recommendation 8: Employment - Short-Term

(print corporate Recommendation Letter on company letterhead paper)

December 20, 20xx

Mr. Adrian Singer
Project Manager
The Conference Organizers Inc.
5700 Orberlin Drive, Suite 200
San Diego, CA, 92121

Dear Mr. Singer:

Juanita Williams worked under my direct supervision for the seven-month period, May through November 20xx. During that period she performed the dual role of Receptionist for the overall Agency, and Administrative Assistant for the Corporate Services Group. Supervising her in those capacities, I believe that I was able to develop a fair idea of Juanita's performance and capabilities, as well as the primary personal attributes which she brought to those jobs.

Although relatively inexperienced when she joined the Agency, Juanita quickly demonstrated that she was a fast-learner and a very willing worker. After a short adjustment period, she performed her primary role as Receptionist in a very efficient and professional manner. At this Agency, that job frequently requires interaction with course participants from many different countries who speak a number of languages, and have different cultural perspectives and attitudes. Juanita was astute at handling these people and situations with tact and diplomacy.

As mentioned, Juanita was also required to perform work as an Administrative Assistant in support of various Institute groups/projects. In this capacity, she always demonstrated that she was a quick learner and very adaptable to performing a wide variety of different tasks efficiently and effectively. She performed extremely well in what was often a multi-tasking situation, working for a number of different supervisors, while still handling her primary Receptionist duties. I was particularly impressed by how quickly and seamlessly she was able to provide support to the Financial/Accounting Group of the Agency, which involved learning the Institute's financial systems and processing varied and complex financial transactions.

On the personal attribute side, Juanita was truly a pleasure to work with. She displayed an extremely friendly, helpful, and outgoing demeanor at all times. As a result, she was very well-liked by both her colleagues and course participants. In fact, I would say that Juanita's pleasant and skilled manner in dealing with the Institute's multi-cultural clientele is likely to generate significant repeat business for the Agency. She has an intuitive understanding of the "client service" and "clients first" attitudes which are so necessary in business these days.

In closing, as I hope the above clearly attests, I would highly recommend Juanita Williams for positions similar to those she held with our organization. In addition, I believe that she demonstrated that she has the potential to perform even more complex and demanding duties.

For further information, please don't hesitate to contact me at (313) 913-1762.

Sincerely,

Mary Steinhart
Director, Administrative Services

Recommendation 9: Employment – Student

(print a Recommendation Letter on corporate letterhead paper)

April 25, 20xx

Ms. Diana Dumbrell
Recreation Director
Pinewood Lake Camp
Pinewood Lake, ON
H3P 4L2

Dear Ms. Dumbrell:

RE: Employment Recommendation - Chelsea Salzberg

Chelsea asked me to write this to you based on our working association over the past three summers. I understand she is applying for a Senior Program Co-ordinator position with you.

I have known Chelsea Salzberg since the summer of 20xx. We first met when she became a volunteer counselor at the City's Pineview Day Camp for Kids. She volunteered that summer since she was not yet old enough to be a paid worker. She did this to both serve her community and to gain valuable counseling experience.

From the very beginning, Chelsea proved to be an eager and highly-motivated worker. During her first year with us she worked with the 5 to 8 year-olds and did a great job of keeping them busy with a variety of imaginative activities and crafts. Many of these group activities were Chelsea's own creation, and I had to spend very little time supervising her once I got her started. The kids loved her too. It was a great first summer.

During the summers of 20xx and 20xx, Chelsea worked at the day camp as a paid counselor. The first two years she was a regular Counselor and in 20xx, she moved on to Senior Counselor duties.

I really can't say enough about how well Chelsea handled her duties those two summers. She was definitely my top performer. She has a special knack for communicating with kids in all age groups. They just seem to relate very well to her and hang on her every word. No doubt, this has a lot to do with Chelsea's unusually friendly and outgoing demeanor. Needless to say, she was very well liked by the kids, her peer counselors, and the City managers.

One particular strength that Chelsea consistently demonstrated was her leadership ability. This became particularly apparent last year when she organized and supervised two off-site field trips. The trips were considered great successes by both participants and staff, and went smoothly.

Overall, Chelsea is a dedicated, hard-working counselor who always puts the needs of the kids before her own. She is a cooperative and willing employee, always ready to go the extra mile.

Accordingly, without hesitation I am pleased to recommend Chelsea Salzberg for the Senior Program Coordinator position at your camp. I have no doubts that she will do an excellent job.
Please call me at 732-1576 if you have any questions.

Sincerely,

Patricia Heywood
Recreation Program Director, City of Ottawa

Recommendation 10: Employment – Manager

(print a Recommendation Letter on corporate letterhead paper)

November 10, 20xx

Mr. Vincent Coulombe
President and General Manager
Westmere Entertainment Corporation
1675 Victoria Ave.
Westmount, QC H2B 3H7

Dear Mr. Coulombe:

<u>Re: Letter of Recommendation - Steven Morrison</u>

Steven Morrison has requested that I write this recommendation letter to you in support of his application to become Entertainment Manager of the IMC Sports Complex.

I have known Steve for more than 20 years, 15 of which he has worked for me as a manager in a couple of my restaurant ventures. Overall I have found Steve to be a hard working and exceptionally productive employee who I can trust.

I first hired Steve back in 1985 to manage a small bistro that I had just purchased at the time in Old Montreal. When I took that business over it was failing. I have no doubt that Steve played an important role in helping to improve the efficiency of that operation to the point where I was able to sell it as a flourishing business four years later. Steve had a lot to do with developing a new menu and streamlining our kitchen and serving operations.

After that, I lost touch with Steve for a couple of years when he moved to Calgary to work in the catering business. However, in 1998 when I decided to open my new steakhouse in downtown Montreal, Steve was one of the first people I contacted to help launch that business. Fortunately he agreed to move back to Montreal to take charge of restaurant operations. Within six months of opening, that restaurant was the busiest steakhouse downtown and it had already earned a reputation for excellent food and impeccable service. Again, most of the credit goes to Steve. He knows the restaurant business inside-out and he is a master at attracting and keeping high quality staff.

Over the years, I have found Steve Morrison to be an exceptional employee and person. He is highly trustworthy and can be trusted to work independently without supervision. He is very likeable fellow and a great communicator, both qualities that make excellent people want to work with him.

In closing; based on knowing and working with him for more than 20 years, I can recommend Steve Morrison to you as entertainment manager of your new complex, without reservation. I have no doubt that he will be able to play a key role in the success of your new venture.

Sincerely,

Pierre Rochon

Recommendation 11: Employment – Sales (Fashion)

(print Job Recommendation Letter on corporate letterhead paper)

May 21, 20xx

Ms. Barbara Lemay
Director, Human Resources
Holt Renfrew Ltd.
1300 Sherbrooke Street West
Montreal, QC H3G 1H9

Dear Ms. Lemay:

Re: Recommendation – Sofia Kindler - Collection Specialist – Giorgio Armani (womens)

I have been asked by Sofia Kindler to provide this recommendation letter in support of her application for the above-noted Collection Specialist vacancy that you are currently advertising on your website.

Sofia has worked under my direct supervision as a Senior Fashion Advisor at Lifestyle Fashions Inc. in downtown Toronto for more than five years. During that time, she has always been regarded as an exceptional and valued employee. In fact, if she wasn't compelled to relocate to Montreal for family reasons, I am sure she would still be with us for many years to come.

During her time with Lifestyle Inc., Sofia has gained diverse experience in retail sales and service, working at a number of our retail outlets in the Greater Toronto Area. She is the kind of results-oriented sales and customer service professional that excels in the retail fashion industry. Sofia was consistently among our top 5% of sales performers, and has twice been named Collection Sales Associate of the Year.

Sofia is a sales professional who prides herself on providing her customers with a superior shopping experience, backed up by personalized customer service. Her knowledge and experience in direct customer sales and service of premium apparel and accessories targeting a diverse group of women is impeccable. In her time with us Sofia has personally developed a loyal client base by providing customers with superior levels of service and satisfaction. She has been involved in all aspects of retail sales and service activities including: direct customer sales and service, merchandising, sales team leadership, and client follow-up.

Sofia's other strengths include abilities to: think and work analytically, identify and set goals/targets and achieve them, participate as member of a sales team, and communicate effectively. In addition to being ideal qualities for a sales associate, these abilities will help her on her long term career track which is to eventually become a fashion buyer.

When Sofia showed me the job posting on your company's website I immediately saw her as an ideal candidate to fill that position. In my opinion, should she be offered the position it will definitely be Toronto's loss and Montreal's gain.

Please feel free to contact me using the coordinates in the letterhead should you need any clarification or if you require any additional information.

Sincerely,

Angela Richler
Fashion Sales Manager

Recommendation 12: Employment - Technical Internship

(print Internship Recommendation Letter on corporate letterhead paper)

March 28, 20xx

Peter Creighton
Director, Creative Services
Pixar Animation Studios
1200 Park Avenue
Emeryville, California 94608

Dear Peter Creighton:

Re: Recommendation - Miranda Kolyn - Technical Director Internship, 3D Modeller

Miranda Kolyn has asked me to be one of her recommenders in her application for the above-noted internship position that you are currently staffing. Last summer Miranda worked for me as a technical assistant and 3D modeller at the Walt Disney Feature Animation Studio in Burbank. During our time working together daily for four months, I found Miranda to be an excellent worker and an exemplary employee. As a graduate student in the Department of Digital Art at Bedford Art College, Miranda is definitely well qualified for your consideration.

In her work as a 3D modelling intern, Miranda was involved in creating complex organic models and sculpting detailed character maquettes. Over the summer I saw her make significant progress in developing her modelling skills. I could see that she has a keen ability to visualize things in a very precise and multi-faceted way, which gives her an excellent sense of three-dimensional design. It was clear that Miranda's real-world experience as a 3D modeller with us, contributing to an actual feature film, intensified her already strong passion and love for the 3D animated feature film industry.

As a person and employee, Miranda was exemplary. She worked very effectively with other team members and demonstrated excellent verbal and written communication skills. She was very well liked by her co-workers and was always willing to help out her peers whenever needed.

Summing up, I think it is fair to say that Miranda Kolyn is among the most gifted modellers I have worked with. I think that her time with Disney last summer just further enhanced and honed her natural 3D modelling abilities. It seems to me that the Technical Director Internship you are offering would be the perfect next step for Miranda. I therefore recommend her for that position, without reservation.

Please don't hesitate to contact me if you would like to discuss this recommendation. I can be reached by E-mail at xxxxxxxxxxx or by phone at xxx-xxx-xxxx.

Sincerely,

Chuck Frommer
Director, Feature Animations

Recommendation 13: Employment - Refusal to Write

(print a Recommendation Letter on corporate letterhead paper)

September 10, 20xx

Melissa Rasmussen
429-A Central Ave.
Cincinatti, OH, 45202

Dear Melissa:

Suzanne Trimble told me that you had called wondering if I would provide you with a letter of recommendation for a new position for which you are applying.

I thank you for the honor, but after some consideration, I'm afraid I will have to decline to write you a letter. Believe me, it is nothing personal. As far as I know, your relatively brief stay with our company was a productive one for you and a pleasant one for your colleagues.

However, since you didn't report to me directly, I don't feel that I know you or your work well enough to be able to make a recommendation one way or another. I would suggest that you request a letter from Angela Goodall since she was your boss for your final six months with the company. Perhaps you have already done so.

Angela would be in a much better position to provide a fair assessment of your work performance during your stay with us. I'm sure she would be agreeable to writing you a letter.

Of course, if you are simply looking for someone to confirm your term of employment here, and your salary level, I suggest you ask Roger Simmons, the Director of Human Resources. I'm sure that his staff could get something like that out to you in a hurry.

I wish you the very best in the future.

Sincerely,

Kenneth Thomas
President and CEO

cc: Angela Goodall, Director of Marketing

COLLEGE-RELATED RECOMMENDATION LETTERS

Recommendation letters are usually required for entry into undergraduate and graduate programs at a college or university. Many graduate programs require two or more letters of recommendation as part of the program admission requirements.

In fact, even if a letter isn't specifically required for an undergraduate program application, it is still a good idea to attach at least one or two. These will make your application stand out from the others.

Normally, these program admission recommendation letters are written at the request of the program applicant by people who are familiar with that person's academic career to-date, and their future education and career aspirations.

For an undergraduate applicant, these letter-writers could include such people as: high school teacher, principal, or vice-principal, community leaders, volunteer program coordinators, clergy, part-time employers.

For graduate program applicants the typical recommendation letter-writer would be academics and professionals who have a solid direct knowledge of the work and character of the applicant: these people would typically be: undergraduate faculty members, administrators, academic supervisors, and professionals in the applicant's discipline.

Of course, it almost goes without saying that whoever writes these letters must be familiar enough with the applicant to be credible, and to be able to answer specific questions about their recommendation if contacted.

Drafting Tips – College-Related Recommendation Letters

Many of the tips for writing employment-related recommendation as well as reference letters also apply to college-related letters. So, please review those sections along with this one.

However, pay close attention to the following tips because they include subtle differences that apply very specifically to college-related recommendation letters. If you are asked by someone to write a college-related letter of recommendation for them, or if you need one written for you, here are some points to keep in mind:

Make Sure You're The One

If you have been asked to write a recommendation letter for someone, it is important that you feel comfortable with the task. If you feel that you don't know the person and/or their academic achievements and aspirations well enough, you should decline, stating that as your reason.

Also, you may find that it would be difficult for you to say very much positive about the person requesting the letter, in which case you should also decline.

Allow Plenty of Lead-Time

If you are the person requesting that someone draft a recommendation letter for you, give that person as much lead-time as you can. It's not fair to ask for a recommendation letter a day or two before it's required. I would say one week as an absolute minimum.

A person asked to write a recommendation letter at the very last minute has every right to refuse. Also, quality will suffer if the letter is rushed at the last minute.

Gather Background Information First

Before starting to write, you should have certain background information available. You'll need to know exactly to whom you are writing. You may also need the requestor to supply you with other relevant background information such as: resume/cv, recent transcripts, copies of academic papers, etc.

In the opening paragraph, list all of the background information such as: your organization name, position title, your relationship to the applicant, when and how you met, time-frame and dates covered, and any other relevant overall background information.

Write To A Specific Person

Recommendation letters for college and university programs should always be personalized. Even if the instructions you receive just provide a mailing address, make a point of finding out the name and title of a specific person to write to. Just check out the Web site of the institution and/or make a phone call.

Personalizing the letter will give you more credibility and will make yours stand out from the others. In other words, do not address one of these letters: "To Whom It May Concern" or "Dear Admissions Secretary."

Get Input From the Requestor

In addition to the basic background information already mentioned, you may need some additional personal information from the requestor/applicant. You might want to ask them to provide you with a checklist summary of key points that they would like you to cover.

You might also want to get input from them on their academic/career aspirations.

Cover the Entire Person

The recommendation letter should address all aspects of the applicant that relate to their ability to perform in the program being applied for. Many colleges and universities provide actual checklists of the categories to be covered in recommendation letters. In fact, research has revealed that there are literally hundreds of different checklists like this, depending on the institution involved.

Following is a list of "typical" attributes and characteristics that would normally be covered in a recommendation letter for a graduate or undergraduate program:

- Intellectual ability
- Knowledge of the field
- Employment and/or academic potential
- Motivation
- Initiative and creativity

- Work habits
- Seriousness of approach
- Emotional makeup and maturity
- Communication skills and abilities
- Adaptability
- Level of commitment, focus

I have even seen attribute checklists that go as far as requiring information on "personal grooming and hygiene" and "appropriate eye contact"! Use your own judgment as to what makes sense in a particular context, and how far you can go.

Be Specific and Give Examples

It will be helpful if your letter rates your applicant relative to their peers. For this you can use relativity phrases such as: average, very good, excellent, or below average. If you can quantify it would also be helpful. For example, "Robert's overall grade point average puts him in the top 5% of the class".

Always try to give specific examples rather than just making open-ended generalizations. For example, "Daphne's ability as a researcher is outstanding as can be seen in her recent paper titled 'From Quantum Physics To Eternity' a copy of which is attached". Or, "Rob's strong communication skills were highlighted when he presented his paper on 'The Psychology of Addiction" at last year's Regional Conference."

Avoid Controversial Statements/Terminology

Do not make statements that you cannot clearly support with facts. For example, statements such as "I believe Frank displays this tendency because he is the product of a dysfunctional background" are clearly not acceptable.

Make sure you avoid using any words or terminology that could be construed as discriminatory such as: race, color, religion, political affiliation, sex, sexual orientation, age, physical appearance, handicaps, marital, or parental status.

Use Active, Powerful Words

Neutral words such as good, nice, satisfactory, fair, reasonable, etc. should be avoided. Use active, descriptive words and terms such as: intelligent, assertive, initiator, self-starter, multi-tasker, motivated, hard-working, cooperative, productive, creative, articulate, leader, communicator, team player, innovative, effective, efficient, honest, dependable, mature, etc.

Review the Final Product

Before signing your letter, do a final careful review. Check all spelling and grammar and make sure the terminology used is appropriate.

Read it out loud to yourself and imagine being the recipient. Is it fair and balanced? Does it truly convey what you believe and want to say about the person who you are recommending? If not, revise it.

SAMPLE TEMPLATES - COLLEGE-RELATED RECOMMENDATION LETTERS

The following pages contain fourteen (14) fully-formatted real-life sample templates of recommendation letters written for actual college-related situations.

Notes Re: Sample Letter Formats

- Although the sample templates in this guide are based on actual situations involving real people, identifying details have been altered to protect privacy.

- All of the sample templates in this guide have been reduced in size slightly to fit the book's page format which allows for page headers and footers.

- The samples use font size of 11 points rather than the 12 points I recommend as ideal. Also, the top/bottom and right/left margins have been adjusted to fit the margins of this book. You may want to readjust them for an actual letter.

Recommendation 1: College – Undergraduate

(print a Recommendation Letter on company letterhead paper)

October 25, 20xx

Ms. Jillian Shepperd
Director of Admissions
Admissions and Recruitment Office
MacDonald University
1449 Dorchester Ave. W., Rm 451
Montreal, QC, H3A 1T4

Dear Ms. Shepperd:

I am very pleased to write this recommendation on behalf of Layla Bell.

Layla has been a student in the accelerated liberal arts program at Holymount High for her entire five years of high school. During that period I have observed her grow into a poised and accomplished young woman. She is an exceptional student with excellent grades resulting from diligent work habits.

Layla has superior interpersonal skills and works equally well independently or in a group setting. She also displays good leadership skills when involved in group projects. She is very well liked and respected by both her peers and her teachers.

Among her many service activities at the school, Layla was a coach of the junior track team for the past two years and she was a member of the senior cross-country team. She also took part in the Mentoring Program and helped a number of juniors navigate their way through their first year of high school. In addition, Layla was involved in organizing a number of fund-raising projects at the school, including a team marathon event that raised over $5,000 for cancer research.

Layla has shown an ongoing interest in world affairs and international development. It is my understanding that she intends to pursue an Honors degree in Political Science or Sociology. She has traveled extensively and has written outstanding reports with observations on conditions she has witnessed throughout the world.

I believe that Layla Bell has tremendous potential as a student and I feel quite confident that she would be an asset to both student life and academics at MacDonald University.

Yours truly,

Allan S. Fenton
Vice-Principal

Recommendation 2: College - Mature Student

(print a Recommendation Letter on company letterhead paper)

December 15, 20xx

Professor Reiner Holzman
Director, Teaching Studies
Boston University Center
115 Beacon St., Rm. 489
Boston, MA, 02110

Dear Professor Holzman:

Re: Recommendation Letter - Deanna Daryna

I have been asked to write this letter for Deanna Daryna as one of the requirements for her admission to the Teaching English As A Second Language (TESL) program.

I have known Deanna for a total of fifteen (15) years. In 1999 we started out together as colleagues in the advertising department of Reader's World Magazine. We worked together there for a period of six years during which I got to know her quite well. Then, in 20xx Deanna moved to our Atlanta office for two years to take part in a special assignment team as the company started to gear-up its Internet presence. When Deanna returned to head office in 20xx she became my employee since I had been promoted to Director of Advertising Operations the previous year.

Accordingly, I know Deanna in a professional capacity primarily as an advertising copywriter and direct marketing expert. What I can tell you about her is that, in those two areas, she is among the very best in the business. She has a gift for copywriting, and her creative mind was responsible for some of the most successful marketing campaigns that Reader's World has ever implemented.

Deanna was hand-picked by senior management to work in Atlanta with the Internet Presence Project Team. As team leader, she headed up a small group responsible for adapting our direct mail campaigns to the Internet environment. The company implemented that team's strategy in early 20xx and now Reader's World is among the top three publishing companies on the Web.

During the dot com meltdown in 20xx, our company went through a major downsizing exercise. Deanna decided that she wanted to take an early retirement package and do something different with her life. I was not at all surprised when she told me that she wanted to go back to school and learn to teach English as a second language. In our early years at Reader's World I had often seen Deanna in the teaching role as she trained new employees from all over the world about how to implement the company's mailing campaigns. I could see that she very much enjoyed that teaching role and how well-suited she was to it. She had a special knack for communicating with employees from our regional offices in developing countries. So, in a sense, Deanna's new path in TESL is a natural progression from her training work with the company.

In my opinion Deanna Daryna is an exceptionally bright and hardworking individual who throws herself enthusiastically into what ever she undertakes. Accordingly, I have no hesitation whatsoever in recommending Deanna as a participant in your TESL program.

Sincerely,

Robert Jamieson
Director, Advertising Operations

Recommendation 3: College – Undergraduate - Job

(print a Recommendation Letter on corporate letterhead paper)

December 1, 20xx

Mr. Roland Zimmerman
General Manager and Editor
Windy City Publishing
Clarkson Building, 12th Flr.
200 East Huron St.
Chicago, IL, 60610

Dear Mr. Zimmerman:

I am pleased to be able to write this letter of recommendation for Valerie Douglas. I understand that she has applied for a job with you as a manuscript reviewer. I have known Valerie since she entered Lakeland College as a freshman three years ago. Over the years I have gotten to know her as a student in two of my classes and also as her department head.

From the very first year, Valerie impressed me in a number of ways. She is a very serious student but can also be very enthusiastic about everything she does. I first met Valerie when she was a student in my introductory English literature class. She immediately struck me as a very earnest person who became increasingly excited about the course material as the semester progressed.

I was very pleased when Valerie asked me to sign her "intent to become an English major" form. Since then, I have watched her use both her seriousness and enthusiasm to develop into an outstanding English major. She possesses an excellent knowledge base in English literature and has developed strong analytical, critiquing, and writing skills.

Recently, Valerie participated in my Early Classics Seminar series, and her performance in the class was excellent. Her contributions included: she wrote three strong papers on topics related to the lives of classics authors, she collaborated with three of her classmates on a skillfully written "period novella", she participated actively in group discussions, and she led a professionally delivered group presentation.

Valerie's strong work ethic, coupled with her meticulous attention to detail, allow her to produce such high quality papers. Her ability to work well with others, even under stress, allows her to get the most out of group assignments. Her strong performance as a group leader was due to rigorous preparation, and her excellent oral communication skills allowed her to highlight how well she knew her subject.

If you are seeking someone who has consistently demonstrated the following skills, abilities and characteristics during her undergraduate career, then I suggest you give Valerie your full consideration.

- Carries out projects successfully
- Works effectively as a team member and handles conflict well
- Shows initiative, creativity, and persistence
- Demonstrates effective time management
- Speaks and writes articulately and persuasively

In summary, I am pleased to recommend Valerie Douglas to you without reservation. If you have any questions regarding the above, please don't hesitate to call me at (312) 335-9747.

Sincerely,

Donald R. Atkinson, Ph.D.
Professor and Department Head

Recommendation 4: College - Graduate - Psychology

(print Recommendation Letter on corporate letterhead paper)

November 20, 20xx

Ms. Catherine Halstat
Assistant Head
Psychology Department
Health Research Center
Midwestern College
1345 College Park Road
Minneapolis, MN, 55450

Dear Ms. Halstat:

This letter is in support of Jason Fairbank's application for admission to your graduate program. Jason is currently a senior psychology major at Midwestern. I have known him since we met in his senior year of high school during a Midwestern familiarization session. We made contact again as soon as he enrolled as a freshman and I have followed his progress ever since.

I have been at Midwestern since 1990 and rarely have I seen a psychology major with as much going for him as Jason has. I believe he has a high potential for success in graduate school.

Jason's greatest strength is his ability to perform research. As a sophomore, he enrolled in Internet-Enabled Research Methods, our senior-level research course. Students in this course must complete a full research project (from literature review to final paper) in only one semester, and they must present the results of their research at the meeting of the Midwest Psychology Research Symposium that is held annually at the University of Southern Illinois.

In my opinion, Jason's presentation was among the top three given at the conference that year. His clarity and composure during the presentation, and his poise and ability to think on his feet during the question period that followed were truly remarkable for a sophomore. He presented another paper at that conference in his junior year which was even better than his sophomore paper. Jason conducts himself very professionally at these conferences and is particularly adept at networking with students and faculty from other institutions to gain more insight into their research and share his own findings.

One of Jason's papers at last year's conference was a presentation of a new student evaluation form that he and several other students developed as a research team. That presentation went over so well that Jason and his team members were asked by the chairperson of the conference to present their research at last spring's National College Conference. They were the first students in Midwestern's history to present at the National Conference. After that, their evaluation methodology was approved by the board and the faculty and has been in use college-wide ever since. This is a good demonstration of Jason's ability to lead a cooperative team effort; a quality that I'm sure will serve him well in your graduate program.

Jason has taken four courses classes from me over the years: Introduction to Psychology, Social Psychology, Developmental Psychology, and The Psychology of Addictions. He has performed exceptionally well in all of them. He is an intelligent, enthusiastic, well-prepared, well-spoken, and assertive individual who gets a lot out of his classes and also gives back a considerable amount through his active participation. His comments in class are always timely, relevant, and valuable in summarizing the discussion or stimulating it to go in new directions. His writing is clear, concise, and articulate.

Jason also has excellent computer skills and his minor is in information technology. He is a self-starter who can work independently with little or no supervision. One particular strength that Jason possesses is the ability to integrate and synthesize information from a variety of sources to produce novel, yet realistic, creative interpretations and conclusions. He used this gift especially well in my Developmental Psychology course in which he scored exceptionally well in his independent project work.
One of Jason's most remarkable traits is his deep-rooted sense of moral and ethical responsibility. We have had many conversations about the ethical climate of both Midwestern College and US universities and

colleges in general. As a result, Jason undertook a major research project that deals with the issue of academic integrity. One of the papers that he presented at last year's National Conference entitled "Dishonesty in Academia – Myth or Reality?" looked at the various types of typical academic dishonest behavior (e.g., cheating and plagiarism) and estimated their degree of prevalence at both Midwestern and at other campuses across the country. Jason was recently invited by the President of this college to present his findings on this at an upcoming Regional College Forum.

As you know, success in graduate school is a function of many variables including intelligence, motivation, communication skills, and personal traits. Although often overlooked in letters of recommendation, this last factor may well be one of the most important at the graduate studies level.

As an individual, Jason is an extremely well-rounded person. He is likeable, straightforward, down-to-earth, confident, and generally, a pleasant person to be around. He is very well-liked and held in high regard by both his peers and members of the faculty. When he has the time, Jason serves as a volunteer mentor in the Freshman Assistance Program and off-campus he is involved in a number community fund-raising projects. Somehow, in the midst of all of these academic and community service activities, he still manages to find the time to play squash. In fact, he was a member of Midwestern's Intercollegiate Squash Team that placed third at the Regional Championships earlier this year.

In conclusion, based on the foregoing it should be clear that I regard Jason Fairbanks as an outstanding individual in every sense of the word. Accordingly, I recommend Jason to you for acceptance into your graduate studies program, without hesitation. I believe he is exactly the type of undergraduate student that I would accept into a graduate program if I were making such a decision. In my opinion, Jason has unlimited potential and will eventually go far as a professional psychologist in the coming years.

Sincerely,

Frederick Wyatt, Ph.D.
Department Head

Recommendation 5: College – Graduate - Business

(print Recommendation Letter on corporate letterhead stationery)

December 20, 20xx

Ronald Lerner, Ph.D.
Associate Dean - Business Administration
Louisiana University
450 College Street
Monroe, LA 71206

Dear Dr. Lerner:

This letter is in support of Winston Bagnell's application for admission to your graduate program. Mr. Bagnell recently graduated from the International Business School in Bad Nauheim where he was a student in three of my classes: Banking and Stock Exchange, Corporate Finance, and International Banking. I have been lecturing at that school for about five years now and rarely have I seen a student with as much potential as Winston. Accordingly, I believe he is an ideal candidate for graduate school.

In my opinion, Mr. Bagnell's greatest strength is his analytical ability. Although he demonstrated this on an ongoing basis, an excellent case-in-point was his performance on his final examinations near the end of his studies. During a four hour exam, Mr. Bagnell and his peers were required to write a detailed analysis about the peculiarities of international merger and acquisition transactions, a fairly complicated subject. The results of Winston's analyses were nothing short of astonishing. Not only did he give very detailed correct answers for each question, but he also approached them from a holistic point of view, displaying true insight into the field of international financial management. Consequently, his paper was awarded the highest possible grade.

In addition, I also observed that Mr. Bagnell possesses excellent communication skills. He consistently demonstrated these capabilities in his in-class presentations. He always presented his subjects convincingly, in a poised and professional manner. On numerous occasions I noted his remarkable ability to think on his feet and maintain his composure when responding to difficult questions and engaging in discussions, both during and after his presentation sessions.

Mr. Bagnell also demonstrated strong leadership abilities in his group project assignments. Particularly notable was his performance in leading his group in a project dealing with "unfriendly takeovers". The lead role he took in coordinating that project, motivating and focusing the other team members, and leading the final presentation were clearly evident.

Overall as a student, Winston performed exceptionally well in each class he took with me. He is an intelligent, enthusiastic, well-prepared, well-spoken, and assertive individual who gets a lot out of his classes and also gives back a considerable amount through his active participation. His comments in class are always timely, relevant, and valuable in summarizing the discussion or stimulating it to go in new directions.

As an individual, Winston Bagnell appears to be an extremely well-rounded person. He is likeable, straightforward, down-to-earth, confident, and generally, a pleasant person to be around. He is very well-liked and held in high regard by both his peers and members of the faculty. He likes to engage in extracurricular activities. For example, he was a key player in the organization of a business conference, and he was also part of the project team that restructured the school's computer network.

In the future, I can see Mr. Bagnell working as a consultant in international corporate finance. His strong analytical, communication, and leadership abilities make him a prime candidate for assignments with major consulting companies that deal with complex international finance issues such as divestitures and mergers and acquisitions. Given the breadth of his skills and capabilities I have no doubt that Mr. Bagnell will be successful in whatever challenges he takes on in his field.

Based on the foregoing, you can see that I regard Winston Bagnell as an outstanding student and individual in every sense of the word. Accordingly, I highly recommend that you accept Mr. Bagnell into your graduate business program. He is exactly the type of undergraduate student that I would be happy to accept into a graduate program if I were making such a decision.

If you have any questions, please don't hesitate to contact me at (213) 472-1950.

Sincerely,

Henry Frenken, Ph.D.
Professor, International Business Studies

Recommendation 6: College – Graduate - Stanford

(print Recommendation Letter on corporate letterhead stationery)

December 20, 20xx

Gilbert Lawrence, Ph.D.
Director of Admissions - MBA
Stanford Graduate School of Business
510 Memorial Drive
Stanford, CA 94215

Dear Dr. Lawrence:

It is with great pleasure that I recommend Jennifer Chang for acceptance into your MBA program.

I have been acquainted with Jennifer since she joined Conrad Technologies Inc. in early 20xx. I have often worked closely with her as her direct supervisor and have found her to be extremely knowledgeable and competent. She is also very enjoyable to work with and is one of the most pleasant persons you'll ever find to work with.

As a person, Jennifer has numerous positive qualities. She is forthright, but neither intimidates nor intrudes. At the same time, she does not easily give up her position on a matter she believes in simply because a different position is held by her supervisors. She's open, friendly, and authentic, and is well-liked by everyone in our office, co-workers and management alike. She often goes out of her way to be friendly and helpful to her colleagues, and people often trust her to take care of their accounts when they are away on vacation or business trips.

As a professional, Jennifer is a resourceful, creative, and solution-oriented person who is frequently able to come up with new and innovative approaches to her assigned projects. She functions well as a team leader when required, and she works effectively as a team member under the direction of other team leaders. From the very beginning, she steadily increased her sales volume. Jennifer also has superior written and oral communication skills.

Throughout her years with us, Jennifer has consistently displayed a remarkable ability to handle frustrating situations with patience, tact, and diplomacy. As an account manager, it is part of her job to deal with and find solutions for the errors of others. No matter how frustrating a situation may be, Jennifer always seems to be able to find an effective solution. She is always very supportive of the person who committed the error, counseling them on how to avoid the same mistake in the future.

I believe that the foregoing people skills are good examples of Jennifer's leadership capabilities. She has an unusual ability to listen to people and understand them. She doesn't hesitate to assume the major responsibility whenever she is in a team situation. She has been an integral part of our marketing team at three different computer trade shows including the largest and most important one in the industry, Comdex Las Vegas. At these events, I found her performance to be exceptional. She always focused on the cohesiveness of the team and demonstrated superior interpersonal and communication skills.
Two hallmarks of Jennifer's character are her honesty and integrity. This extends through her interpersonal and business relationships, to her integrity of thought. Her thinking process is scientific, thorough, and

meticulous, and she approaches analytical tasks with an objective and exacting eye. This is what I mean by integrity of thought.

The only area of weakness that I have ever noted about Jennifer is her lack of formal business training. She acknowledged this weakness when she approached me about writing this letter. Our discussion on the subject concluded that the best course of action is for her to pursue further studies in business. This is what led to her candidacy for your MBA program.

As noted above, I have seen Jennifer develop into a mature and sophisticated professional who is now ready to move on to the next level. Accordingly, she has set her sights on obtaining more advanced education in business, which she hopes to receive at your prestigious university. I understand and support this capable young woman's aspirations and wish to give her all the support I can.

I therefore recommend Jennifer Chang to you with enthusiasm, and I would greatly appreciate it if you would favorably consider her application to the Stanford University MBA Program.

If you have any questions please do not hesitate to contact me at (202) 232-4690.

Yours sincerely,

Charles Conrad
President and CEO

Recommendation 7: College - MBA Program - Haas

(print Recommendation Letter on corporate letterhead stationery)

March 15, 20xx

Director of Admissions
Haas School of Business
Evening & Weekend MBA Program
University of California at Berkeley
1450 University Way
Berkeley, CA 93105

Dear Director of Admissions:

<u>**Recommendation: Peter J. Kendall**</u>

I am very pleased to write this recommendation letter in support of Mr. Peter Kendall's application for admission to the Haas School of Business MBA Program, Fall Term 20xx.

Mr. Kendall is one of eight consultants that I have personally supervised for the last three years in my capacity as manager for the PeopleSoft, Inc., Corporate Sales Support, CRM group. He was hired by me in late 20xx and has been under my direct supervision ever since. During that period, I had the great pleasure of watching him blossom from an IT database expert at the beginning, into a fully functioning business-oriented professional Sales Application Consultant.

Peter Kendall is a hard-working self-starter who invariably understands exactly what a project is all about from the outset, and how to get it done quickly and effectively. One thing in particular that impresses me is the way in which he works cooperatively and communicates effectively with others while working on the projects to which he is assigned These projects usually require close liaison with other parts of the company including developers, sales product consultants, product strategy staff, and others. Peter has consistently demonstrated a remarkable ability to interact with all project personnel in such a way that everything is well understood by everyone involved and the project remains on track. I cannot remember an instance in which he did not meet his project objectives.

I observed Mr. Kendall as a resourceful, creative, and solution-oriented person who is frequently able to come up with new and innovative approaches to his assigned projects. He functions well as a project manager, and he also works effectively as a team member under the direction of other project managers when required. He is a fun-loving individual who is well-liked and respected by his co-workers, and he is always ready to extend a helping hand to others.

Peter always seems to stay one step ahead of everybody else on the new technology front. He is the only member of my team who has been certified in Customer Relationship Management by PeopleSoft , when this certification is not required for his job function. In addition to his technical expertise, over the years Peter has sharpened his business knowledge and skills and has developed an excellent ability to perceive the big picture. For example, he has a good understanding of how new product features can be used to strengthen the marketing message for a new campaign.

In summary, I believe that Peter Kendall possesses a rare blend of superior analytical, technical, and interpersonal skills. His unique combination of tenacity, analytical abilities, and good business sense is a mix that one doesn't find every day. Accordingly, I can enthusiastically recommend Mr. Kendall as a prime candidate for your business school. I believe he has excellent potential to become a high-caliber senior manager in the future and a credit to the business school which he attends.

For clarification or further elaboration, please feel free to contact me at (925) 694-4455.

Sincerely,

Kevin Glass
Manager, Sales Support

Recommendation 8: College - Graduate IT Program

(print Recommendation Letter on corporate letterhead stationery)

March 5, 20xx

Director of Admissions
Faculty of Commerce
University of Wollongong,
NSW, 2522, Australia

Dear Director of Admissions:

<u>Recommendation Letter – Henry K. Lam</u>

I am writing this letter for Henry Lam as one of the requirements for his admission to your Master of Industry - Information Technology Program.

I have known Henry for a total of six years. In 20xx we started out together as colleagues in the Information Technology department at NETNow Information Technology Limited. We worked together there for a period of 18 months during which I got to know him quite well. Then in 20xx, Henry and I, and a former classmate started our own business: iDea2000 Internetworking Solutions Provider.

In a professional capacity, I know Henry primarily as an expert in Internet Service Provider Systems and Networking. In my estimation, he is among the very best in those two areas Henry has a gift for Internet-based business, and his creative mind was responsible for some of the most successful marketing campaigns that iDea2000 has ever implemented.

As CEO of iDea2000, Henry spearheaded our special campaign to provide Web hosting and Internet total solutions to customers in the Internet environment. The company implemented that strategy in early 20xx, and now iDea2000 is among the top ten internetworking solution providers in Hong Kong. In spite of the dot com meltdown in 20xx we have managed to maintain an average annual revenue growth rate of 20%. The company started with three employees and now has a workforce of 10. Under Henry's leadership, iDea2000 has followed a strategy of gradual and balanced growth which has carried us smoothly through the recent upheavals in the worldwide Internet industry.

In 20xx the company will be vigorously pursuing a new business development opportunity that has been identified and Henry will be at the helm. It involves the creation of an online trading analysis Web site. We expect a significant downturn in the worldwide stock market due to the uncertainty brought about by the imminent war between the United States and Iraq. Based on historical trends, once that situation has been resolved or stabilized, stock markets will undergo a rapid and significant recovery.

That upturn, coupled with the Hong Kong Exchange's launch of its third generation automatic order matching and execution system (AMS/3), will result in Internet-based trading services becoming increasingly active. Other factors, including   Hong Kong Exchange's impending cancellation of the minimum agent commission scheme, will heat-up the stock market activity even more. We are convinced that with all of this activity going on, investors will be clamoring for a Web-based comprehensive and customizable stock analysis interface focusing on the Hong Kong market.
We believe that time is of the essence, and timing is critical if we are to going to leverage this window of opportunity. That is why we have made this project our all-consuming business development focus for the

entire year. In order to fast-track this into a successful business development venture we know that our company needs additional technical and business knowledge. We believe that sponsoring Henry Lam on your program to acquire this additional knowledge is the prudent strategic choice for our company at this time.

In closing, I believe that Henry Lam is an exceptionally bright and hardworking individual who throws himself enthusiastically into what ever he undertakes. Accordingly, I have no hesitation whatsoever in recommending Henry as a participant in your Master of Industry - Information Technology Program on behalf of our company. I believe that, given his demonstrated skills, abilities and experience, Henry's participation will prove mutually beneficial for both our company and the university.

Sincerely,

Jenny Lee, M.B.A.
Senior Vice-President

Recommendation 9: College – Master of Journalism

(print Recommendation Letter on corporate letterhead stationery)

July 16, 20xx

Roger Tinsdale
Director, MJ Program
School of Journalism
Carleton University
St. Peter's Building 255
1175 Colonel By Drive
Ottawa, ON K1S 5B6

Dear Roger Tinsdale:

<u>**Letter of Recommendation – Linda Hubert**</u>

I am writing this letter of recommendation at the request of Linda Hubert, in support of her application for the Master of Journalism program, at Carleton's School of Journalism and Communication.

Please let me qualify first by stating that this is more of a personal reference than a professional one since, although I have known Linda for many years, I have never worked with her directly in a professional capacity. That being said, I have kept in fairly close touch with her over the years, so I have a pretty good idea as to what she has been up to on the professional front.

In fact, I have known Linda Hubert since her days studying Journalism at Carleton University some 30 plus years ago. I recall that after she graduated with her undergraduate degree in Journalism she went on to hold a series of progressively more senior positions as copywriter and editor for various government departments and agencies in Ottawa, and ultimately became senior editor for a number of industry and professional publications.

Having read quite a few of Linda's written products in those days, I always admired her skill as a copywriter. She has the ability to break a complex technical subject down and write about it in plain, understandable language. Along with that, she has an excellent capacity to elicit information from technical experts and specialists and then translate that into highly readable and understandable copy. Linda also has excellent oral communication skills, a talent that I know she has enhanced greatly in recent years while teaching and leading workshops and seminars internationally.

On the personal attributes side, Linda is a quick-study and once she decides to do something she pursues it with a passion. She is also very bright and can quickly adapt to new material and situations. She also works well independently, a trait which served her well over the years as she free-lanced and pursued various entrepreneurial projects.

After reviewing the MJ course summary/outline material on your website I believe that Linda would be an excellent participant in your MJ program. In fact, I think that Linda's early background in Journalism and copywriting/editing, coupled with her many years of unique life and professional experience, would add value to both the MJ course and the participant group. I therefore am pleased to endorse Linda Hubert's participation in Carleton's Master of Journalism program.

Please feel free to contact me at (514) 798-1965 if you require additional information.

Sincerely,

Robert Carruthers, M.B.A.

Recommendation 10: College - MBA Program (UK)

(print Recommendation Letter on personal or institutional letterhead paper)

215 North Street
St Andrews, KY16 9AJ

December 20, 20xx

Postgraduate Admissions Office
University Office
University of Aberdeen
King's College
Aberdeen, AB24 3FX

Dear Admissions Director:

I have known and worked with Miss Samantha Wu for almost five years and based on that I am delighted to recommend her for your MBA programme.

I am currently a postgraduate student at St. Andrews University in the United Kingdom. For the past 20 months, I have had the pleasure of working with Samantha on the Mustard Seed Project, an educational project for school children in remote areas of South East Asia and China. As her project partner, I have seen Samantha demonstrate outstanding leadership in motivating others to participate in the project as well as through her efforts in initiating and operating the project. Below I share my assessments of some of Samantha's key attributes.

I first met Samantha in Yunnan Province of China in 20xx, while we were working on a documentary film project. We discovered that we both cared deeply about human rights matters and discussed many educational issues. In 20xx, Kimberley sent me a proposal to help her set up a project to assist school children in remote areas have better access to educational resources and change the way they see the outside world. She believed that many school children in remote areas of South East Asia and China ran away from home to the outside world with a dream of escaping from poverty but instead ended up involved in prostitution or drug dealing. The plan was to encourage travelling backpackers to share their real-life knowledge of the world outside with school children in those remote areas while the backpackers themselves could benefit from opportunities to explore the local cultures. Samantha's ideas and passion convinced me, and I was the first person to get involved in the project and later became her primary project partner. Later on, other post graduate students also joined the project.

Compared with the leaders and participants of other non-profit organizations or projects in which I have participated, such as the Irish Agriculture Farmers project and the Xuannan Primary School Volunteer project (China), Samantha stands out in a number of ways.

1. Persistency in initiating the project
   Over a three year period Samantha actively participated in various non-profit organizations so that she could learn from their operations. During the Mustard Seed Project initiation, many of us thought the project scope was too small. In response, Samantha called several meetings to discuss the scope, where she shared her experiences in similar small non-profit projects. After a few follow-up emails, the team united and reached a shared goal that: in three years we would have volunteers teaching programs in three schools. She then led us through the detailed planning tasks for achieving that goal. Step by step, she taught us to pursue our vision in a realistic way. Samantha also showed respect for the ambitious ideas of team members by placing our previous big plans in different project phases. Her persistence in building the team was the foundation that allowed us to move forward.

2. Discipline in managing the project
   Compared with other community projects in which I have been involved, Samantha operates the Mustard Seed Project team very efficiently and keeps us on track at all times. She sends out regular email reminders to advise us of our progress and she insists on meetings of the core team once every

two weeks to discuss the issues and current tasks, even when she is on business trips. She also sends out decision records and follow-up assignments after each meeting. In a small scope non-profit project where everyone has only limited time to devote, her management skills keep the progress of the project on schedule.

3. Flexibility in internal and external communications
    Samantha demonstrates superior communication skills while dealing with people from a variety of backgrounds including; project team members (mostly post-graduate students), local partners who are local small businessmen, and volunteer backpackers. As the project leader, she listens to and respects our opinions and always makes sure we fully understand the reason behind her decisions and are comfortable with them. When building partnerships with local communities for the Mustard Seed Project, she always first listens to the opinions of the locals about educational issues facing school children, and only after that does she explain the design of our courses and the mission of our project. She sees the project from the perspectives of both the project team and the local community. She knows how to successfully earn the trust of local community groups first. When dealing with project volunteers, Samantha always addresses the benefits from both the local school children's angle and also from the perspectives of volunteers. She emphasizes the unique local cultures and environments that will enrich the volunteers' travel experience.

As the founder and the leader of the Mustard Seed Project, Samantha has had a major impact. In the past 20 months she has led us to many achievements including: completing the teaching materials, building connections with local communities, completion of one volunteer teaching session in Pang Numtrang School in Thailand, and building a partnership with a local businessman there to advertise our project in his hostel. However, I believe her impact on this project goes beyond just making it happen or operating it. Samantha's passionate but down-to-earth personality has influenced us not to just dream about changing, but also to find, plan, and execute the changes. This is one of the main strengths she offers the non-profit world -- a pragmatic approach backed up by concrete and proven methods.

Overall, I think Samantha Wu is a person with big dreams and serious plans who has the ability to achieve them. I believe that by studying at your institution, she will get closer to her dream of making an impact on the world.

Based on the foregoing, and without reservation, I am pleased to highly recommend Samantha Wu for acceptance into your programme. If you have any questions, please do not hesitate to contact me.

Sincerely,

Jill Wong
St. Andrews University, UK
School of Social Anthropology

Recommendation 11: College - Ph.D. Candidate

(print Recommendation Letter on corporate letterhead stationery)

March 5, 20xx

Dr. Milton Quigley
Associate Dean
School of Advanced Psychology
University of Wisconsin Graduate School
Milwaukee, WI 52501-0340

Dear Dean Quigley:

<u>**Re: Regina Halldon**</u>

It is a pleasure for me to be able to write a letter of recommendation supporting Regina Halldon's application for admission to your Ph.D. program in psychology. I have taught psychology at St. Francis University for 17 years and Regina is one of the most outstanding undergraduate students that I have ever encountered.

Regina Halldon earned a total grade point average of 3.85 out of a possible 4.0 counting all of her psychology courses over three years as an undergraduate. This is one of the highest grade averages achieved in our undergraduate phychology program in recent years. In her final year Ms. Halldon published an exceptional paper on Jungian Psychology in the student journal, *PsychedIn.*

Her final paper was based on Kant's *Critique of Pure Reason*, wherein Kant argues, largely in response to Hume, that the mind brings an elaborate a priori structure to its interactions with the world. This is Kant's transcendental psychology which contains arguments for a number of theses about the nature of a subject that has knowledge of objects. Patricia Kitcher and Peter Strawson have radically different positions on how to interpret Kant's Subjective Deduction and the transcendental psychology contained therein. Strawson rejects Kant's transcendental psychology completely; Kitcher advocates a limited acceptance of it. Ms. Halldon's paper presents Kant's transcendental psychology and argues forcefully against a number of Kitcher's and Strawson's criticisms. I have urged her to rewrite the paper and submit it to a good journal.

Finally, Regina Halldon is a personable, teachable, hard working and highly intelligent student. She is extremely articulate and always stands out among her peers for the clarity and cogency of her arguments. Many of my students make good use of email but the emails from Ms. Halldon were about the most interesting communications that I have ever had from a student. I have rarely had a student who acts so quickly and positively to address critical comments and suggestions.

In short, Regina Halldon is one of those rare students who clearly has the potential to make a difference. I could not recommend her more highly.

Yours sincerely,

Randall Hurtibise, Ph.D.
Professor of Psychology

Recommendation 12: College – Special Award

(print a Recommendation Letter on corporate letterhead paper)

August 31, 20xx

Professor Wayne Bradshaw
Chairman - Selection Committee
CSUN Outstanding Graduate Award
Administration Building, 3rd Floor
Woodland Hills, CA 91367, USA

Dear Professor Bradshaw:

The purpose of this brief letter of recommendation is to support the nomination of Annie Klein for this year's Outstanding Graduate Award. I have known Ms. Klein for the past four years in my role as her faculty advisor.

I believe that Annie is an outstanding example of someone who has struggled tenaciously and managed to overcome considerable adversity in life to become an exceptional student and an exemplary member of both the CSUN community and the San Fernando community at large.

After learning in 20xx that her international academic qualifications were not acceptable and that she would have to start college studies from the beginning, she decided to do whatever had to be done to fast-track an Honors BA. Consequently, through a personal regime of commitment, hard work and dedication to excellence, she completed her BA program at CSUN in the winter of 20xx with a GPA of 3.95. Essentially, she achieved in 2 ½ years what takes most students four years or more.

Although she was functioning as a single mother, fast-tracking an Honors BA, and volunteering at the SFPD (San Fernando Police Department) doing administrative work and dealing with victims.-- all at the same time -- Annie also went out of her way to tutor and counsel a number of her fellow students who were having difficulties with their studies. In addition to all of that, she interned at the MTA (Metropolitan Transportation Authority) for a six month period while pursuing her studies at CSUN.

I believe that Annie's personal story is an exceptional testament to how a person can overcome unusual hardships and daunting challenges in life, achieve her goals and ambitions, and truly realize her dreams. She grew up in poverty in a war-torn country and then immigrated to the USA in her early twenties to start her adult life here from scratch. A few years later she found herself the divorced single mother of an infant, destitute and living on welfare; shunned by her family and community.

Since then, over just a few years, I have watched Annie Klein lift herself from the bottom of the heap to a position of success and respect in her chosen field while at the same time managing to excel in her studies. In addition she is a committed single mom, raising a bright little boy.

In closing, in view of Annie Klein's many accomplishments in both her academic and personal lives over the past few years, I believe that she would be a highly deserving recipient of this year's Outstanding Graduate Award.

Please let me know if you need additional information.

Sincerely,

Marion Greenstreet
Academic Advisory Services

Recommendation 13: College - Scholarship Applicant

(print a Recommendation Letter on corporate letterhead paper)

August 31, 20xx

Margaret A. Smeltzer
Chairperson
Scholarship Selection Committee
Pacific Coast University
2705 Valleyview Drive, Rm. 205
San Diego, CA, 92120

Dear Mrs. Smeltzer:

I am writing this letter in support of Daniel Kim's nomination for the 20xx-08 California Human Resources Society Scholarship.

I have known Daniel since he entered PCU as a freshman three years ago. Since that time I have come to know him very well as his teacher in four of his courses, and through various academic and extracurricular interactions. I am writing this because it is my sincere belief that he is eminently qualified as a deserving recipient of your scholarship. Over the past three years he has demonstrated to me that he possesses exceptionally high levels of academic ability, motivation, and potential for contributions to the field of human resources management.

Daniel earned a grade of "A" in all four courses that he took from me: Psychology - An Introduction, The Science of Psychology, Developmental Psychology, and Research Studies in Psychology. Please note that the criterion for an "A" in all four of these classes is only attained by receiving 95% of all possible points. The Science of Psychology is an extremely rigorous course, and fewer than 2% of students have ever received an "A" in this subject in the 15 years that I have been teaching it. His performance in Developmental Psychology was just as remarkable. He received perfect scores on 10 of 12 tests and earned 98 out of a possible of 100 on his term paper.

Daniel's current accumulated GPA is 3.87 (out of 4.0) and his GPA in psychology classes is 4.0. He has never received a grade lower than an "A" since his freshman year. He will graduate "with Honors" next year. This superior level of academic performance indicates an exceptional understanding of his subject material. With knowledge, skills, and abilities at this level, Daniel should go far in his field in the future. I believe he plans to attend graduate school, and we are currently looking at his options in this area.

One of Daniel's key strengths is his ability to analyze, integrate, and synthesize information from a wide variety of disparate sources and to apply the essence of this information to an important topic. His term paper, titled "The History of Cognitive Dissonance and Its Prevalence In The Corporate Workplace" is an excellent example of this ability. A copy of that paper is attached.

Students writing papers were asked to accomplish the following three objectives:

- Document the history of a particular area of psychology in which you have a particular interest.

- Explain a controversial issue that continues to exist within this area.

- Take and defend a specific position on this controversial issue.

As you will see if you have the time to even skim the attached, Daniel's paper is meticulously well-researched, argued, and written. It was by far the best in the class. His work was also very well organized and his writing style was simple, clear, and precise. His discussions were thoughtful and the rationale for his arguments were well laid-out. His research was exceedingly thorough, and he cited some 38 sources although only 15 were required.

Daniel's paper is the ideal precursor for a major research project in human resource management that he plans to undertake next year on Morals and Ethics In the Workplace. That course requires the completion of an original empirical research project with the aid of computer applications software at each stage (i.e., bibliographic search, statistical analysis, graphics production, and word processing). The final requirement of this course is the presentation of the completed paper at the annual Regional Undergraduate Research Conference. I have absolutely no doubt that Daniel will produce a research paper that reflects positively on himself, our department, and PCU.

Daniel is the perfect example of an exceptional student whose actions speak louder that his words. Because of his academic prowess and his achievements to-date, in the PCU Psychology Department, Daniel "walks softly but carries a big stick", as the expression goes. He is highly esteemed by his peers and the faculty members. Both groups are aware that he is a young man of considerable talent, intelligence, and motivation who knows how to go the extra mile and contribute his skills to the group in a quiet, professional manner.

As an active and committed member of the PCU Psychology Club, Daniel has set a new standard for club activities. His leadership abilities have not gone unnoticed. This past year he was elected National Secretary of the American Association of Undergraduate Psychology Clubs (AAUPC). He has also been re-elected President of his local chapter for next year.

As teachers, every once in a few years we may be fortunate enough to teach someone who we believe has what it takes to go on and make a major difference in their field. I believe that Daniel is one of those rare individuals.

In closing, without hesitation I would like to recommend that Daniel Kim be chosen as the recipient of the California Human Resources Society Scholarship, for 20xx-xx. This young man deserves it.

Sincerely,

Mathew Houston, Ph.D.
Professor and Department Head

attachment

Recommendation 14: To Recommender – Dental

(print Recommendation Letter on personal letterhead paper)

375 Highplains St., No. 5
Albuquerque, NM 87120

June 5, 20xx

Professor Andrew Reardon, PhD
Department of Biology
University of New Mexico
Albuquerque, NM 87131

Dear Professor Reardon:

Thank you very much for agreeing to write a letter of recommendation in support of my applications to dental schools. The purpose of this letter is to give you some background information about me and to transmit to you some support documents that you will need.

As I mentioned to you, I am currently in the process of applying for dental school. I have found that the material that you covered in your course is excellent preparation for the upper-division biology courses. Most notably, the Bio210A lab class has steered my interest towards biology and helped me in my work as a research assistant at the UNM Langstaff Biomedical Research facility. Between work and volunteering I have managed to maintain good academic standing, graduating this year with a bachelor's degree in Biochemistry and Cell Biology.

My passion for science and my belief in the importance of health care are the main factors that have motivated me to want to become a health care provider. However, my desire to pursue a career as a dentist stemmed primarily from my own personal experiences. Having gone through many extensive dental treatments myself, I am well aware of the significance of dentistry in the healthcare spectrum and believe that there is a need to educate others about the necessity of good dental care through outreach.

During the past year, I volunteered as a dental assistant and X-ray technician for the UNM Student-Run Free Dental Clinic, and I also worked at a dental office in Alta Monte. Recently, I had the opportunity to volunteer with the Wishes Become Smiles Foundation in Ensenada, Mexico, which specializes in cleft palate reconstruction, orthodontic treatment, and provides general dental care for the underserved population of Baja California. Through these experiences I have realized that research and clinical outreach can significantly reduce the burden of dental-related diseases. At the same time, I have learned that education about good dental hygiene can improve people's oral health and overall well-being. As a member of the UNM Pre-Dental Society, I am proud to be the Dental Health Education Coordinator, where I hope to continue my duties in community outreach and health education.

Once again, I very much appreciate you taking the time to write a letter of recommendation on my behalf. I am targeting June 30th to make my submissions to dental schools, so I will need the letter before that date. I have provided an envelope for you to mail the letter directly to the UNM Career Services Center if you wish. However, I would prefer to pick up the sealed letter from you so that I can thank you in person.

If there is any additional information that you need, please don't hesitate to contact me via phone at xxx-xxx-xxxx or by email at xxxxxxxxxx. Your support means a great deal to me!

Yours sincerely,

Kara Tan

Encl

SAMPLE TEMPLATES – GENERAL RECOMMENDATION LETTERS

The following pages contain five (5) real-life templates of recommendation letters written for various general situations not specifically covered earlier.

Notes Re: Sample Letter Formats

- Although the sample templates in this guide are based on actual situations involving real people, identifying details have been altered to protect privacy.

- All of the sample templates in this guide have been reduced in size slightly to fit the book's page format which allows for page headers and footers.

- The samples use font size of 11 points rather than the 12 points I recommend as ideal. Also, the top/bottom and right/left margins have been adjusted to fit the margins of this book. You may want to readjust them for an actual letter.

Recommendation 1: General – Court Appointment

(print a Recommendation Letter on corporate letterhead paper)

October 20, 20xx

Honorable Robert Dickson
U.S. Bankruptcy Court
795 Oakmount Street, 5th Floor
San Francisco, CA 94104

Dear Judge Dickson:

I am writing this letter in support of one of my parishoners, Mr. Ashleigh Gardner, CPA, in his recent application to be named as a Bankruptcy Trustee for San Francisco County. (See application letter attached). It is my understanding that you may be in a position to put in a good word on his behalf to the U.S. Trustee's Office.

I have known Ashleigh Gardner for more than five years as both a parishioner in this Diocese and through his services to the Episcopal Church and community in general. I believe him to be highly trustworthy and dependable person as well as a skilled and competent professional in his field.

Mr. Gardner has been involved extensively in the life of the Diocese both personally and professionally. One valuable service he has provided over the years has been his provision of specialized tax advice to each newly appointed priest. In addition, he and his associates routinely provide diocese staff with accounting, payroll, tax, and general accounting and financial advisory services. These services have all been supplied by his company to the diocese on a pro bono basis.

Recently, Ashleigh Gardner and his associates were called upon to do a special audit of the more than 30 churches in the Diocese. As a further testament to his esteemed standing in the parish, Mr. Gardner has served for 5 years as the Trustee of St. Mark's Episcopal Church on Franklin Ave., San Francisco.

In light of my personal knowledge of Mr. Gardner, and the valuable and exemplary service he has given to this Diocese, I strongly support his application for Bankruptcy Trustee. I believe that any additional support that you could give to Mr. Gardner's application would ultimately be of significant benefit to both the Diocese and the community at large.

Please do not hesitate to contact me should you wish to discuss this further.

Yours faithfully,

Stevenson Haydon
Archbishop

Encl.

Recommendation 2: General – Community Work

(print a Recommendation Letter from a private citizen on regular stationery)

98 North Ridge Road
Minneapolis, MN, 55420

June 30, 20xx

Dr. Alan Banting
Director, Adult Counselling Services
Holy Cross Memorial Hospital
453 South 6th Street – Admin. Bldg.
Minneapolis, MN, 55425

Dear Dr. Banting:

Thank you for your letter asking me for my assessment of Fred Smalling as possible member of your Adult Counseling Advisory Committee. It is my understanding that Fred provided my name to you as a possible reference.

I have known Fred for a little over 20 years, both as a neighbor and a fellow volunteer on many community projects. Shortly after we moved into the neighborhood in the early 1980s, I joined the Friendship Program at the hospital and discovered that the chairperson of that committee lived right across the street from me. It was Fred. Since then, we have had fairly regular contact.

Although we don't socialize that much as neighbors, I know Fred best through his work with the Police Association Assistance Program and through his involvement with the Friends of Schizophrenics group.

Based on my volunteer work with Fred over the past two decades, I am pleased to be able to recommend him as a member of your new committee. He is a dedicated, community-minded individual who puts helping others less fortunate than himself as his number one priority. One thing in particular that always amazed me about Fred is his ability to leave his stressful job with the police force at the end of the day and then freely give up two or three evenings of every week to the various causes he supports. Many weekends too. He's clearly an exemplary citizen.

Fred has an amazing ability to relate to people from all walks of life. He is also very smart and has excellent organizational and leadership abilities. Fred is almost always the one who takes charge and motivates both clients and volunteers. Having been so involved in community service work over the years, he is very knowledgeable about fund-raising techniques, and he has many contacts in his network. You might want to consider him seriously as overall Chair, or at least, Fund Raising Chair, on your committee.

In closing, I'm sure that Fred Smalling would be a valuable addition to you committee. I only hope that his involvement with that won't take him away from all of the good work that he already does for so many. You can reach me at 543-5678 if you have any questions.

Yours in service,

Bradley Finnegan

Recommendation 3: General – Corporate

(print a corporate Recommendation Letter on corporate letterhead stationery)

November 10, 20xx

Ms. Deborah Weekes
President and CEO
Primal Fashions Inc.
790 Westwood Beach Plaza
Suite 2015
Los Angeles, CA 90095

Dear Deborah Weekes:

Recommendation of Quantum Solutions Inc.

I have been asked to write this letter of recommendation to you by Paul Stevenson, CEO of Quantum Solutions Inc.. It is my understanding that your company is considering engaging Quantum to upgrade your corporate information systems.

From March to October 20xx, Quantum Solutions implemented a comprehensive IT monitoring system at Barnesworth LLC, where I am CFO. That solution involved the installation of a number of major software modules as follows: IBM Tivoli Monitoring, IBM Tivoli Monitoring for Databases, IBM Tivoli Monitoring for Messaging and Collaboration, IBM Tivoli Enterprise Console, and IBM Tivoli Netview.

We chose Quantum because they offered a complete turnkey solution which included the ability to personalize and customize the monitoring software and associated rules. Their monitoring solution has turned out to be a resounding success, providing us with capabilities beyond our original expectations. We now have a much improved overview of our entire IT infrastructure, which allows us to more effectively optimize performance and prevent failures.

Quantum Solution's in-depth knowledge of IT operations allowed their team to complete the project on-time and within budget. In particular, Kevin Chapman and Anthony Maurissmo of Quantum, were found to be resourceful, creative, and solution-oriented professionals. From day-one, and throughout the project, we could see that they were truly committed to the success of the installation.

Accordingly, without hesitation I can highly recommend the Quantum Solutions Team to undertake any project in their field of expertise in the belief that it will be completed in a timely, cost-effective, and professional manner.

Yours sincerely,

Stephen Waites
Chief Financial Officer

Recommendation 4: General - Special Award

(print Recommendation Letter on corporate letterhead or personal stationery, as appropriate)

May 25, 20xx

Susan Alison
Program Director
Maine Exceptional Children Foundation
750 Harlowe Street
Bangor, ME 04401

Dear Susan Alison:

The purpose of this letter is to recommend that Mr. Anton Weldon be named as recipient of the 20xx Michael F. Hall Teacher of the Year Award, as sponsored by the Maine Exceptional Children Foundation.

In my opinion, Mr. Weldon consistently provides exceptional service to students with disabilities and their families. He continually demonstrates through his actions that he is an outstanding special education teacher and administrator. Without a doubt, Anton has had a positive impact on the education and quality of life of children and youth with exceptionalities and he has enhanced their interactions with their families. In the seven years I have known him, Anton has always gone above and beyond what was expected of him. He constantly puts the students first and provides them with opportunities to grow.

In addition to his excellent teaching and program development skills, Mr. Weldon has also demonstrated considerable administration and fiscal management abilities. For example, when districts were faced with challenging budget cuts, Anton and his special education director developed a program streamlining proposal and presented it to the Eastern District School Board. He had managed to find a novel way to involve not-for-profit community groups in program delivery to the special needs community in the district. That Community In Action Program was an instant success, and it continues to function today.

Mr. Weldon's coursework development is second to none. The special education program that he created for the district is in its third year of implementation and has been recognized by parents, staff, and neighboring districts as innovative and exemplary. The success of the program is, in large part, due to the methods that Anton uses when presenting the material and how effectively he engages students in the learning process. In fact, engaging students is one of Anton's great strengths; one that enables him to produce some amazing results. He is a major booster of the Special Olympics and he often coaches teams in volleyball, basketball, archery, bowling and track.

Anton also works tirelessly in the community to promote his kids. He works hard to make them employable by leveraging his contacts throughout the community where he has developed a network of personal commitments and he makes sure that all of his students have direct access to all community mental health programs. Anton is highly respected by his peers and co-workers for his selfless commitment to helping young people with special needs.

In my opinion, Anton Weber is an amazing human being and is a gift to all the lives he touches. Accordingly, I believe he would be a very worthy recipient of the 20xx Michael F. Hall Teacher of the Year Award.

Sincerely,

Director, Special Education Programs
Eastern Maine District

Recommendation 5: General - Travel Visa Application

(print corporate Visa Recommendation Letter on corporate letterhead)

April xx, 20xx

Canadian Visa Office
400 Water Street
Los Angeles, CA 91101

Dear Visa Officer:

I am writing this recommendation letter on behalf of Mr. Sheldon Wong, our Senior Developer and Program Manager. Mr. Wong is planning to attend the Canadian Institute's Annual Forum on Anti-Money Laundering in Toronto from April 28 to April 29 as a representative of CCC (California Collections Services, Inc.). Accordingly, CCS assumes all financial responsibility for any debts that may be incurred by Mr. Wong during his brief stay in Canada.

CCS is currently helping a growing number of top lenders that offer credit card, real estate, student, and installment loan products, to manage their charged-off, delinquent, and pre-delinquent portfolio operations. Our potential clients include BMO (Bank of Montreal) in Canada. The information and contacts acquired by Mr. Wong in this meeting will help our company build an improved collections system that will better protect our client organizations from fraud as well as other financial risks such as money laundering.

We are actively sponsoring Mr. Wong's U.S. green card application which will allow him to work permanently in the United States. We look forward to his continued contribution in building a better risk management system for CCS.

I thank you in advance for your cooperation and assistance in the timely processing of Mr. Wong's visa application so that he may attend the above-mentioned forum on our behalf.

Should you have any questions, please do not hesitate to contact me by e-mail at: xxxxxxxxxx, or you can call me directly at xxx-xxx-xxxx.

Very sincerely,

Leonard Ranford
CEO, California Collection Services, Inc.

REFERENCE LETTER TEMPLATES

These are more general letters that are often requested by employees when they leave the employ of an organization. Normally factual in nature, they are usually addressed, "to whom it may concern" and provide basic information such as: work history, dates of employment, positions held, academic credentials, etc.

Reference letters sometimes contain a general statement (as long as a positive one can be made) about the employee's work record with the company that they are leaving. Employees often submit these letters with job applications in the hope that the letter will reflect favorably on their chances for the new position.

Character reference letters are sometimes required by employers when hiring individuals to perform personal or residential services such as child care, domestic services, etc. These letters are usually drafted by a former employer and deal with such characteristics as honesty, dependability, and work ethic/performance.

TYPES OF REFERENCE LETTERS

There are four primary categories or types of reference letters as follows:

- Employment
- College
- Character
- General

Employment-Related Reference Letters

These are normally specifically requested by the person the letter is being written about, often when they are leaving a job.

These letters are typically more general in nature than employment-related recommendation letters and are usually addressed "To Whom It May Concern" or "Dear Sir/Madam". They tend to state facts to confirm a person's employment such as position title, duration of employment, etc.

Nevertheless, as explained earlier in this guide, some people and organizations use the terms reference letter and recommendation letter synonymously (see page 9). In those cases, what they call reference letters may be more specific, and may be addressed to a particular individual. However, this is the exception, not the rule.

College-Related Reference Letters

These are typically used for college program admission and/or to recommend individuals for scholarships. Normally, they will be more generic than recommendation letters, and will often be addressed "To Whom It May Concern" or, to an office using salutations such as: "Dear Admissions Director" or "Dear Admissions Committee".

Character Reference Letters

These reference letters usually deal with a person's general personality characteristics and attributes such as: honesty, integrity, level of commitment, work ethic, values, community service, empathy, etc. These letters too are normally addressed "To Whom It May Concern" but can be written to a specific person in some situations.

General Reference Letters

General reference letters cover all other types of general references. These are always addressed in a general way using "To Whom It May Concern" or "Dear Sir/Madam". Typically, they deal with such issues as credit references, bank references, customer references, business references, and the like.

WRITING REFERENCE LETTERS

The information in this section is very similar to the writing tips listed earlier under employment- and college-related recommendation letters. Nevertheless, there are some subtle differences specific to reference letters. So, review them carefully.

As discussed earlier, I choose to repeat some of the information here that was covered earlier so that this could become a stand-alone section and you won't have to flip back-and-forth and get even more confused between reference letters and recommendation letters. (It's already confusing enough, isn't it!)

At the risk of repeating myself on this point, please remember that, although a reference letter is one "type" of letter of recommendation, it is somewhat different from the recommendation letters that we covered earlier.

Again, reference letters are much more general in nature and are usually addressed "To Whom It May Concern". Whereas, recommendation letters are more personalized and detailed and should, almost without exception, be addressed to a specific person.

Drafting Tips – Reference Letters

If you are asked by someone to write a "letter of reference" for them, here are a few important points to keep in mind:

Make Sure You Are the Right One

If you feel that you don't know the person well enough or are not the appropriate person you should decline. This is not as critical a situation as it is for recommendation letters since your reference letter will be more general in nature and not directed to a specific person.

However, remember that your personal "John Henry" will be in the signature block.

Typically, if an employment-related reference letter, you would be asked to write it by someone who is leaving your organization to take another job. Normally, you would have a certain moral obligation to supply such a letter.

But watch out! I recently read an article stating that lawyers for some big firms are advising their corporate clients not to commit themselves to paper in general reference letters for fear of later legal repercussions in cases where they might give a good reference and then later the employee somehow screws up.

To me, this is paranoid thinking. But we all know what lawyers are like when it comes to covering rear ends (no offense meant!). My advice is; just make sure that whatever you write is honest and fair.

Start With the Background Parameters

As mentioned above, for a reference letter you will normally not be writing to any particular person. It's usually addressed "To whom it may concern". In the opening paragraph provide all of the background information such as: your relationship to the person, organization name, position titles, time-frame and dates covered by your assessment, and any other relevant background information.

In fact, for most reference letters you shouldn't have to get much more specific.

Get Additional Input From Requestor

If you feel you need more information, don't hesitate to ask the requestor for a copy of their resume/cv. In addition, if you have access to them, you may want to have a quick review of recent performance evaluations.

You can also ask the requestor to jot down some key points they would like you to mention if possible (at your discretion). This would include things like highlighting their work and accomplishments on a special project and/or their participation on a special task force.

Don't Get Too Specific

Since a reference letter is somewhat general and open-ended in nature, don't go overboard with specific traits and details about the person you are writing about. It will be helpful to use meaningful job-performance-related terms and words, although in a much more general sense than one would in a recommendation letter.

If it's employment-related, your reference letter should address all or most of the areas of the person that are job-performance-related, but in a more general sense. Specific examples are less necessary than in a regular recommendation letter.

For example, such statements as "I observed Wendy to be a hard-working, highly-effective team participant, with strong communication skills" would suffice.

Avoid Controversial Terminology

As with recommendation letters, make sure you avoid using any words or terms that could be construed as discriminatory: race, color, religion, politics, sex, sexual orientation, age, physical appearance, handicaps, marital or parental status, etc.

Use Active, Powerful Words

Avoid neutral words such as good, nice, satisfactory, fair, etc. Use active, descriptive terms such as: intelligent, assertive, initiator, self-starter, motivated, cooperative, productive, hard-working, creative, articulate, leader, communicator, team-player, innovative, effective, efficient, honest, dependable, mature, etc.

Review the Final Product

Before signing your letter, do a final careful review. Check all spelling and grammar and make sure the terminology used is appropriate. Read it out loud to yourself and imagine being the recipient. Is it fair and balanced? Does it truly convey what you believe and want to say about the person? If not, revise it.

SAMPLE TEMPLATES – EMPLOYMENT-RELATED REFERENCE LETTERS

The following pages contain twenty (20) real-life templates of reference letters written for various employment-related situations.

Notes Re: Sample Letter Formats

- Although the sample templates in this guide are based on actual situations involving real people, identifying details have been altered to protect privacy.

- All of the sample templates in this guide have been reduced in size slightly to fit the book's page format which allows for page headers and footers.

- The samples use font size of 11 points rather than the 12 points I recommend as ideal. Also, the top/bottom and right/left margins have been adjusted to fit the margins of this book. You may want to readjust them for an actual letter.

Reference 1: Employment – Sales Position (insurance)

(print an Employment Reference Letter on company letterhead stationery)

September 25, 20xx

[Recipient address- Line 1]
[Recipient address- Line 2]
[Recipient address- Line 3]
[Recipient address- Line 4]

To Whom It May Concern:

RE: Employment Reference – Tony Rocco

This is to confirm that Tony Rocco worked under my direct supervision as an Insurance Sales Representative from January 1999 to October 20xx.

During this period Tony progressed from sales initiation trainee to fully certified sales representative by the time he left for another job.

I would say that Tony is a hard working individual who learns quickly. He is generally cooperative and can perform well as a member of a team, although he prefers to work independently. He communicates very well orally and is working hard to improve his writing skills. I saw considerable improvement in this area during the last year he was with us.

As a sales professional, Tony was always above the 50th percentile of performers on my sales team. Since he started from scratch only two years ago, and most of his colleagues were highly experienced with strong, established networks, I would say that Tony achieved a lot in his early years as a sales representative.

So, based on what I observed during my two years with Tony, I believe he has excellent potential to become a high-performing sales professional in the insurance business.

For further information, I can be reached at (514) 989-0779.

Sincerely,

Jannik Soberman
Director, Insurance Sales Operations

Reference 2: Employment – Sales Position (pharma)

(print Reference Letter on corporate letterhead paper)

August 28, 20xx

[Recipient address- Line 1]
[Recipient address- Line 2]
[Recipient address- Line 3]
[Recipient address- Line 4]

To Whom It May Concern:

<u>**Re: Ashleigh Hawthorne**</u>

Ashleigh Hawthorne was a member of my pharmaceutical sales team for a little over two years between June 20xx and September 20xx. During that time I believe I gained a good idea of his character and capabilities as an employee, as follows.

Ashleigh was an above average sales person, consistently meeting and sometimes exceeding his targets. Pharmaceutical sales can sometimes be very difficult due to intense competition and the high knowledge level of the medical professionals. Coupled with these factors, the sales representatives must work alone and only meet as a group twice monthly. Ashleigh dealt with these factors better than most and was among my top three performing sales professionals during the two years he worked for me.

He also demonstrated above-average administrative capabilities, consistently submitting his sales and trip reports on time and completed accurately. He communicated well both in-person and in writing. On a few occasions I attended sales presentations made by Ashleigh, and was each time impressed by his oral presentation skills. At sales meetings he was a primary contributor and displayed leadership qualities.

Although I didn't socialize with Ashleigh I can say that his personal demeanor was pleasant and he was well-liked by both clients and co-workers.

Based on my impressions of Ashleigh Hawthorne during the two years he reported to me I would not hesitate to recommend him as a valuable addition to any sales team.

Sincerely,

Dennis Rayburn
Manager, Sales and Marketing

Reference 3: Employment – Sales Manager (IT)

(print Reference Letter on corporate letterhead paper)

December 20, 20xx

[Recipient address- Line 1]
[Recipient address- Line 2]
[Recipient address- Line 3]
[Recipient address- Line 4]

Dear Sir/Madam,

I have known Richard Quenton for a total of five years; from early 20xx to late 20xx during which time he worked directly for me as a member of our New Business Sales Team.

During that period I saw Mr. Quenton evolve from sales support specialist to Sales Manager for the North of England and the Midlands. I believe that rapid progression is a testament to Richard's exceptional creativity and strong analytical skills which make him stand out from his peers.

From his first day on the team, Richard displayed a remarkable sense of maturity that belied the fact that he was the youngest member of the group. He demonstrated exceptional, initiative, creativity and leadership when he spearheaded a successful campaign to sell the Hamilton Terminal to what was considered an unconventional client base.

Richard is a person of high personal integrity who displays an infectious enthusiasm for his job. He has an excellent sense of humor and was well-liked and respected by his peers, colleagues working under him, and clients alike. His gregarious personality, coupled with his excellent speaking skills, made him a popular presenter at our weekly sales meetings. He is also a hardworking self-starter who understands the concept of working as part of a team.

In addition to his superior analytical skills, Richard frequently went out of his way to help members of his team and contributed to the overall success of the department by always being willing to step-up and cover client meetings and lead sales campaigns and client seminars. His ability to get along with people from diverse backgrounds is exceptional, and resulted in the growth of existing accounts, as well as the steady development of a portfolio of new clients. He was named Salesperson of the Month twice in 20xx.

The only minor weakness I ever noted in Richard was a tendency to be chronically late for group meetings in his early days with the organization. However, as a result of his conscious efforts to improve in this area, by the time he left the company he had completely overcome this problem.

In closing, I have no hesitation in recommending Richard Quenton for any position for which he is qualified and I am certain he will be an asset to your organization. If you have questions or require further elaboration please do not hesitate to contact me at + 44 25 5790 905 791.

Sincerely,

Frank Schecter
Team Leader – New Business Sales

Reference 4: Employment - MIS Manager

(print Reference Letter on company letterhead paper)

October 12, 20xx

[Recipient address- Line 1]
[Recipient address- Line 2]
[Recipient address- Line 3]
[Recipient address- Line 4]

To Whom It May Concern:

Re : Employment Reference – Neil Shankman

Neil Shankman worked under my direct supervision for the eight-month period, from Jan. 20xx to Aug. 20xx. During that time he performed the role of Manager MIS and later he provided advice as a Senior Consultant. Supervising him in those capacities, I believe that I was able to gain a fair idea of Neil's performance, capabilities, and the primary personal attributes which he brought to those jobs.

Neil is a hard-working self-starter who invariably understands exactly what a project is all about from the outset and how to get it done quickly and effectively. During my tenure as a Director at AstroCorp I cannot remember an instance in which Neil missed a deadline. In fact, he often brought projects in ahead of schedule, which allowed him to complete extra unplanned work at times.

Neil is a resourceful, creative and solution-oriented person who is frequently able to come up with new and innovative approaches to his assigned projects. He functioned well as team leader when required, and he worked effectively as team member under the direction of other team leaders.

On the technical and engineering side, he also produced excellent results, even though he much preferred creative assignments. Nevertheless, I think it is fair to say that Neil never let his creative passion get in the way of his technical obligation.

In closing, as detailed above, based on my eight months working with him, I am pleased to recommend Neil Shankman for whatever the position he may qualify for at your company. I believe he is the type of employee that any organization would be happy to have in their workforce.

Sincerely,

Steven Farr
General Manager, Operations

Reference 5: Employment - Student

(print Reference Letter on corporate letterhead paper)

November 5, 20xx

Margaret Whitehead
Senior Editor
Educational Vistas Magazine
3300 South Wabash Ave., Suite 3505
Chicago, IL 60602

Dear Ms Whitehead:

<u>**Re: Employment Reference – Susan Charles**</u>

I am providing this reference letter at the request of Susan Charles, who has asked me to serve as a reference on her behalf. It is my understanding that you are considering Susan for a position on your editorial team as a Researcher.

Please be advised that the information contained in this letter is confidential and should be treated as such. The information I provide should not be disclosed to Susan Charles or anyone in your organization not involved in the hiring decision related to her. Additionally, the information should not be disclosed to anyone outside of your organization without Susan's consent.

I have known Susan Charles for the past three years. During that time she has taken the following courses which I teach: Introduction to Education Studies, Educational Psychology, and Teaching Secondary English. As her professor, I have had numerous opportunities to observe Susan's participation and interaction in class and to evaluate her knowledge level. Accordingly, I would rate her overall performance in these subjects as slightly above average. This is evidenced by her marks of B+ across the board in the courses which I taught.

I can recall Susan making an exceptional presentation on the subject of "different learning modalities" in the Educational Psychology course. The multi-media approach she used for that was original and highly effective in supporting her thesis. In general, the papers she submitted to me were well researched and original in perspective. However, I found that her writing style and approach often left something to be desired. In fact, I can recall pointing out to her on a number of occasions that if her papers were better organized and more succinctly written, her marks would have improved from a B/B+ to an A-/A.

Overall, based on her academic performance in my courses, and my understanding of the Researcher position for which she is applying, I believe that Susan Charles would perform an adequate job from the outset and would likely perform at an above-average level once she has grown into the job, especially if encouraged to work on improving her writing.

If you would like to discuss this further, please feel free to contact me at (312) 852-4936.

Sincerely,

Robert Jamieson, Ph.D.

Reference 6: Employment - Marketing

(print Reference Letter on corporate letterhead paper)

August 23, 20xx

John Hudson
Human Resources Manager
Quantum Technologies Ltd.
750 West Putnam Avenue
Greenwich, Connecticut 06831

Dear John Hudson:

Reference: Donald J. Kemp

I am writing this letter of reference at the request of Donald Kemp. I understand that Donald has applied for the position of Marketing Coordinator with your company.

Donald worked for QTL for a period of approximately three years. During that time he reported directly to me while he occupied the position, Marketing Analyst, Corporate Markets. I believe that my close working relationship with Donald during that period puts me in a good position to evaluate his job performance.

As Marketing Analyst, Corporate, Donald was solely responsible for independently initiating and conducting research and related studies leading to the identification and development of new products and services for our company. After assuming the position he quickly made it clear that he knew what he was doing. Within eight months he had already made two successful product development proposals to our senior management group. Both of these programs were approved and have since become successful additions to our line of software applications for business. The following two years saw more of the same with Donald being personally responsible for the development of three more key products.

Donald is a self-directed researcher cum analyst who works extremely well independently. He has an excellent understanding of the software market for small and medium business. He also possesses superior writing and presentation skills which helped considerably in the development and presentation of new product proposals. Although he prefers to work alone, was required to function as a team member on two special projects on which I observed him to be a contributing and collegial team member.

In conclusion, based on my understanding of your Marketing Co-ordinator position, as described in the job profile that Donald gave me, I would say that he is well-suited to the job. In fact, such a position would seem to be the next logical step in his career development in the marketing field.

Please feel free to contact me if you require any additional information.

Yours sincerely,

Edgar Martinez
Director, Marketing Programs

Reference 7: Employment – Plant Engineer

(print Reference Letter on corporate letterhead paper)

June 21, 20xx

[Recipient address- Line 1]
[Recipient address- Line 2]
[Recipient address- Line 3]
[Recipient address- Line 4]

Dear Sir or Madam:

<u>**Letter of Reference for Mr. Raynald Doerner**</u>

I have been asked to provide this letter of reference by the subject, Mr. Raynald Doerner. Our company is in the process of reengineering its operations and, much to my regret Mr. Doerner's department and position have been eliminated as we streamline our manufacturing operations.

Raynald has been an engineering specialist at Romtech's main processing plants since 1992. He is an expert in stress analysis and the use of analytical and design software such as ADSYS, GtStrent, STEAD, and COMSAR for industrial applications including: structural steel, foundation structures, and piping and pressure vessels. In addition to his technical expertise, Raynald has developed and implemented a number of innovative design approaches such as tension field "stressed skin" cold boxes which reduce or eliminate the need for diagonal box bracing. This design enhancement has significantly reduced the costs of our air circulation housing modules.

Raynald possesses the relatively unique ability to apply analysis to fabrication documents using AUTOCAD software. His design drawings are well drafted with many allowances for ease of construction and thus reduced costs. He consistently produces highly accurate construction documents and calculations. His attention to the smallest details is impeccable.

Overall, Raynald is a highly skilled, conscientious and capable senior engineer. Accordingly, I strongly recommend Raynald for any civil, structural, mechanical or piping engineering position where technical expertise, analytical skills, speed and accuracy are primary considerations.

If you require any additional information, please feel free to contact me at the numbers below.

Sincerely,

Mason B. Berends, PE
Product Design Manager, Plant Equipment

Reference 8: Employment - Software Support

(print business-related Reference Letter on corporate letterhead stationery)

September 25, 20xx

[Recipient address- Line 1]
[Recipient address- Line 2]
[Recipient address- Line 3]
[Recipient address- Line 4]

To Whom It May Concern:

<u>Letter of Reference for Ron Jackson</u>

I am pleased to respond to Ron Jackson's request for a letter of reference.

Ron Jackson worked for my firm, McNaughton, O'Brien, Peyton and Associates, Inc. (MOPA) for the period April 20xx to August 20xx. The firm provides a method for hedging institutional portfolios against downside risk. Its principal clients are pension funds and large investment management firms; its product is delivered through a time-sharing system using MOPA's proprietary software.

Ron Jackson's primary responsibility was the operation and development of our software modules. Those tasks were made particularly difficult by the fact that our programs are written in an older, not widely understood computer language known as APL . In addition the programs were not properly documented. Starting from scratch and on his own, Ron quickly studied and learned about our product and the software that supports it. He showed outstanding diligence in responding to our client's requests for simulations and other applications of the programs. Over the 14 months he was with us he was able to attain a high level of expertise in programming APL which allowed him to make critical changes to the code in order to accommodate many of our clients. He also managed to document our software in such a way that others could easily take over his duties whenever he left.

In short, I believe Ron Jackson has the talent, willingness and ability to be a highly productive staff member in the area of software design, development and application. He has a pleasing personality and I expect he will continue to be successful in future work.

Sincerely,

Manfred Peyton
Director, MOPA

Reference 9: Employment – IT Technical Specialist

(print business-related Reference Letter on corporate letterhead stationery)

July 30, 20xx

[Recipient address- Line 1]
[Recipient address- Line 2]
[Recipient address- Line 3]
[Recipient address- Line 4]

<u>**Reference: Patrick Richter**</u>

I am very pleased to write this letter of reference on behalf of Patrick Richter. In my capacity as CFO of Vistaview, Inc., I have known Patrick as one of my key employees for almost four years. Vistaview is a Web Services development company that specializes in designing, developing and maintaining corporate intelligence databanks. Overall, I believe Patrick to be a skilled, thoughtful, and thoroughly professional Information Technology expert, which I will briefly elaborate on below.

I first got to know Patrick when he helped us develop a software system as a Research Assistant in Computer Science Department at the University of Seattle. On that project I was impressed by his initiative and his thirst for knowledge. He also showed strong analytical and problem solving skills. In fact, it was for those qualities that we hired him at Vistaview, Inc. after his graduation in 20xx.

During the more than three years that Patrick has worked with me at Vistaview Inc., I have always been impressed by his outstanding diligence and his high level of technical expertise. Indeed, I have been more than satisfied with our working relationship. During this period, I also had the great pleasure of watching him blossom from a junior software developer into a fully functioning business-oriented Principal Software Architect who played a primary architecture and software management role on some of our company's key projects. I was particularly impressed by the professionalism and technical innovation that Richard demonstrated on our Corporate Intelligence project.

Not only is Patrick highly intelligent; he also is very hard-working. He has tackled every project with enthusiasm and competence. He invariably understands exactly what a project is all about from the outset, and how to get it done quickly and effectively. He consistently produces superior quality work, on time, and within budget targets. I see Patrick as a true example of the type of employee that every employer wants: accountable and responsible, with a superior work ethic and a high degree of integrity.

In closing, I recommend Patrick Richter very highly for any position in the IT field that involves complexity and creativity, with high quality output requirements. He is a bright and hardworking person who will do his very best to do an outstanding job for whomever he works. If I may be of further assistance with regard to Patrick, please contact me at 402 795-6345 or wdawson@useattle.edu.

Sincerely,

William Dawson, Ph.D.
Professor of Finance, University of Seattle
CEO, Vistaview, Inc.

Reference 10: Employment - Financial Specialist

(print business-related Reference Letter on corporate letterhead stationery)

June 14, 20xx

[Recipient address- Line 1]
[Recipient address- Line 2]
[Recipient address- Line 3]
[Recipient address- Line 4]

To Whom It May Concern:

Reference for Luiz Gonzalez While at Liberty Financial Services, Inc.

Mr. Luiz Gonzalez worked for me at Liberty Financial (Las Vegas) in our financial analysis department from April 1999 through February 20xx. As a Vice-President and head of the analysis department, I had several people performing analytical tasks, but always assigned Mr. Gonzalez the more critical and complex financial structures.

He was especially skilled in linear programming with which he developed several mathematical models that allowed Liberty to provide its major clients (Fortune 500 companies, along with leading domestic and foreign financial institutions) with optimum financial structures for various capital equipment transactions. His pricing techniques allowed our clients to either receive a higher after-tax yield on their investments, or to win transactions through more competitive bidding. Among his many accomplishments were pricing models for "cross border" leases which would allow Japanese corporate investors to take advantage of the tax laws in Japan and the US to provide lower cost financing to US companies.

Mr. Gonzalez was also excellent in training other members of the department and in explaining complex and often very theoretical techniques to our clients.

I was very unhappy when Mr. Gonzalez decided to leave Liberty Financial, but I understood his desire to work on the development of quantitative financial models in the university research environment to which he moved. I was extremely pleased with his general performance, as well as his skills in working and communicating with clients and other employees.

I would highly recommend the hiring of Mr. Luiz Gonzalez. If any additional information is required of me, I can be reached by telephone at (415) 435-1983 or by e-mail at gpratt@liberty.com.

Sincerely,

George Pratt
Vice-President, Financial Analysis

Reference 11: Employment - Financial Controller

(print business-related Reference Letter on corporate letterhead stationery)

November 20, 20xx

[Recipient address- Line 1]
[Recipient address- Line 2]
[Recipient address- Line 3]
[Recipient address- Line 4]

To Whom It May Concern:

Re: Employment Reference – Farad H. Chawdra

I am pleased to be able to write this letter of reference on behalf of Farad Chawdra who worked for me in the position of Controller with my company Market Direct, LLC (formerly Online Financial Group Inc.) since April 20xx.

Farad Chawdra was one of a handful of employees that were key to the success of our company over the past few years. He was responsible for managing a wide range of financial and operational activities of the company. As Controller, he was the senior manager responsible for all aspects of financial accounting, management and control. In addition, his operational responsibilities included: human resources management, payroll administration, loan servicing, IT operations, mergers and acquisitions, and development of offshore operations.

First and foremost, I would describe Farad as a very hard working multi-tasker who always gets the job done. He is a highly intelligent individual who was a valued member of our senior management team. In addition, Farad is an excellent leader-by-example, and he has the ability and willingness to effectively teach others.

In summary, I regard Farad Chawdra as a highly committed and conscientious professional who produces results. I can therefore recommend him without hesitation for employment in similar positions in his field.

If you require additional information, please feel to contact me by phone at 202-754-5952, or by e-mail at rgujurat@onlinefg.com.

Sincerely,

Radek Mortensen
President

Reference 12: Employment - Customer Services

(print business-related Reference Letter on corporate letterhead stationery)

May 23, 20xx

[Recipient address- Line 1]
[Recipient address- Line 2]
[Recipient address- Line 3]
[Recipient address- Line 4]

To Whom It May Concern:

Cedric Gundy – Reference Letter

I have been asked by Cedric Gundy to write this reference letter since he has been forced to move out of State for family reasons and will therefore have to leave his position with ResConn Inc.

Cedric Gundy has worked at ResConn Corporation since July 1999. At the beginning of this year, he was promoted to Senior Financial Officer from Research Support Analyst. I am his immediate supervisor and one of two owners of the company. Resconn provides specialized programming software to structure major capital equipment lease financings. Our clients are primarily financial analysts working at the banks which invest in these transactions. Our day-to-day operations resemble those of a consulting firm since the use of complex mathematical models requires extensive customer support.

Cedric assists our clients in creating equation models to reflect their tax, accounting and economic requirements. He developed our most complex deal structure to-date (a "Shared Allocation Partnership" model) and explained it in detail to a major client. He continues to work with a tax attorney in developing enhancements to Shared Allocation Partnership models which we expect will generate favorable publicity for our products and capabilities with new clients.

By combining his background in econometrics, operations research and financial modeling with a willingness to learn, Cedric quickly became our most technically proficient customer support person. Our clients are especially pleased with his patience and his easily understood explanations. He takes great pleasure in helping them learn mathematical, economic and financial concepts.

Cedric also had the primary responsibility for training and developing two Financial Research Analysts who joined our company in 20xx. It was mainly through his efforts that they both quickly became productive members of our staff. He is now training a third Research Analyst.

I highly recommend Cedric Gundy for whichever position he applies for. Please don't hesitate to contact me should you require further information.

Sincerely,

Gordon Cameron
Vice President, Customer Support Operations

Reference 13: Employment – Research Specialist

(print business-related Reference Letter on corporate letterhead stationery)

October 28, 20xx

[Recipient address- Line 1]
[Recipient address- Line 2]
[Recipient address- Line 3]
[Recipient address- Line 4]

To Whom It May Concern

<u>**RE: Marietta Chang**</u>

I have known Marietta Chang and worked with her in various capacities since 1995. Over this eight year period Marietta has provided me with research and analytical support, financial advice, and clear, concise explanations of complex econometric and statistical relationships. I have always been impressed by Marietta's high level of technical and mathematical knowledge and am more than satisfied with our working relationship.

In addition to unraveling difficult economic, statistical, financial and mathematical problems, Marietta has the rare ability to break down complex problems and situations into simpler components and explain them in terms that non-technical people can understand. This ability is a very valuable asset when it comes to dealing with our clients and staff alike.

I was very sorry to hear that Marietta was leaving us so that her husband could take a new job on the East Coast. She will be very difficult to replace.

In summary, Marietta Chang will be a valuable employee for whichever company hires her to perform functions in her areas of expertise.

Sincerely,

Frederick Constantino
Vice-President, Market Research

Reference 14: Employment - Professional Consultant

(print professional Reference Letter on corporate letterhead stationery)

October 18, 20xx

[Recipient address- Line 1]
[Recipient address- Line 2]
[Recipient address- Line 3]
[Recipient address- Line 4]

To Whom It May Concern:

Letter of Reference - Arun Uday

I got to know Arun Uday over an 18 month period while he worked for me as a consultant on a project sponsored by the World Bank in Jaipur. During that time I was able to closely observe Arun on a day-to-day basis. I am therefore pleased to offer my observations on him as a professional.

Arun Uday joined Thomson Publications under my Directorship as Manager, MIS in early, 20xx. He was later re-designated as Senior Consultant as a result of his superior performance. His specific duties during that period included the following:

- Plan, organize, direct, control and evaluate the operations of information systems.
- Develop and implement policies and procedures for computer systems development.
- Manage a team of IT professionals to design, develop, implement, operate and administer computer networking software, network management software, and information systems.
- Perform network design for multiple locations, including RIP, OSPF and BGP protocols.
- Design and develop the WAN with links between nineteen geographically dispersed locations.
- Design data backups, Internet procedures, and online data transportation mechanisms.

Although inexperienced in the beginning, Arun quickly demonstrated that he was a fast and very willing worker. In a relatively short period he made the adjustment to his role and duties as MIS Manager. He performed extremely well in what was often a multi-tasking situation that required working for different organizational components while still handling his primary IT duties.

I was particularly impressed by how quickly and seamlessly Arun was able to provide support to the financial /accounting group of the organization, which involved learning the project financial structure and the processing of complex and varied financial transactions. He played a lead role in designing the structure of the network that connects the districts to Head Office. He also designed and developed a unique project monitoring matrix.

In summary, based on our time working together, I would rate Arun Uday Sharma as an intelligent, well-organized and hard working professional who brings a lot of enthusiasm to his work. I therefore recommend him highly for future assignments in his field.

Sincerely,

V.J. Haryana
Managing Director

Reference 15: Employment - Professional Services

(print business-related Reference Letter on corporate letterhead stationery)

August 23, 20xx

[Recipient address- Line 1]
[Recipient address- Line 2]
[Recipient address- Line 3]
[Recipient address- Line 4]

Dear Sir or Madam:

I have known and worked closely with Ellen Wei since 1997. During that time, I have developed the highest esteem for her, both personally and professionally, and it is therefore my pleasure to write this letter on her behalf.

I first met Ellen while doing research for my book, *Financially Speaking: Your Personal Guide To Investment For Your Future* - which, I am happy to report, became a national bestseller reaching number four on the *New York Times* business paperback bestseller list in 20xx. I had come across an excellent article Ellen had written on the subject of deciding whether to buy or lease one's car. At that time, I was in the process of writing a section of the book on buying versus leasing, and her expertise proved invaluable to me in formulating my own position on the subject. Ellen then decided to design a special computer program designed to help consumers make sense of the lease/buy dilemma for both houses and cars. I found this software so impressive that I have included Ellen's name and address in the book, so that interested readers can order the program from her directly.

Throughout the book-writing process, Ellen proved herself an indispensable resource, fact-checking the math in the book and catching computational errors that several other experts had missed. She has since continued to prove her worth long after the book's publication. For example, when I was scheduled to appear on a national television show a few weeks ago, Ellen was kind enough to double-check the veracity of the facts and figures I would be quoting. Even though she was quite busy with her own work, Ellen put in long hours to make sure that all of my calculations were correct.

I cannot speak highly enough of Ellen's skills and capacities. In addition to being a brilliant analyst, she is a wonderful teacher - attentive, patient, and thoughtful. She possesses a broad and comprehensive command of mathematical and financial matters, and complements that expertise with a rare ability to elucidate even the most complex problems.

Ellen Wei is an extraordinary thinker, instructor, and friend. She has my gratitude and my respect, and I recommend her services wholeheartedly.

Sincerely,

Sandra Bethel

Reference 16: Employment – Security Services

(print corporate Reference Letter on corporate letterhead stationery)

October 25, 20xx

[Recipient address- Line 1]
[Recipient address- Line 2]
[Recipient address- Line 3]
[Recipient address- Line 4]

To Whom It May Concern:

Employment Reference - Roberto Gregory

I am writing this letter of reference at the request of Mr. Roberto Gregory who was in charge of security services, reporting to me, at Cityscape Hotel for the past five years.

I regard Roberto Gregory as a highly experienced professional in his field. He designed and set up the entire security operation at this hotel before it opened in 20xx. Since that time he has managed all aspects of security operations here, which has resulted in this hotel earning the distinction of being ranked by a leading industry magazine as having one of the best security records in the entire industry.

Something about Roberto that has always impressed my a great deal is his degree of preparedness. He is constantly looking at "what if" scenarios and preparing contingency plans to respond to those situations should they occur. He is the mastermind behind Cityscape's Emergency Preparedness Plan, which is considered to be among the very best plans of its type in the hotel industry in this country.

I also regard Roberto as an excellent member of our management team. He invariably deals with situations in the appropriate fashion, always minimizing the hotel's exposure. In addition, he is a great people person who has the ability to attract, motivate, and keep high quality staff. He always keeps abreast of the latest developments in the field of security and safety.

In light of the above, I have no hesitation in recommending Roberto Gregory for employment in any type of management in the security and safety business, especially in the hospitality industry.

Please contact me at (416) 654-2598 if you require additional information.

Sincerely,

Philip del Fuenta
Hotel General Manager

Reference 17: Employment - High School Teacher

(print teaching-related Reference Letter on school letterhead stationery)

June 28, 20xx

[Recipient address- Line 1]
[Recipient address- Line 2]
[Recipient address- Line 3]
[Recipient address- Line 4]

To Whom It May Concern:

<u>**Teaching Reference – Mario Suarez**</u>

Mario Suarez taught mathematics at Mountain Ridge High School during the second semester of the 20xx-xx school year. He took over from another teacher and on very short notice did a credible job of teaching classes in Calculus, Computer Math, Trigonometry, and Algebra. In total, he taught these subjects to six different classes. He showed a good grasp of the material and was able to develop excellent rapport with his students. He was one of the most popular teachers.

Unfortunately, due to declining enrollment there is a surplus of tenured teachers at our school for the coming school year and we will not be able to rehire Mario. We will truly miss him.

Mario is a high energy self-starter who quickly assumes responsibility and is not afraid to face new challenges and situations. He is also very personable. Not only did he get along well with his students but also with the faculty. He was quick to fit in with an "older" staff and was readily accepted by them.

As Vice-Principal, I would very definitely rehire Mario Suarez given the opportunity. I recommend him highly for your consideration.

Sincerely,

Joan Clark
Vice-Principal

Reference 18: Employment - Student Teacher (favorable)

(print teaching-related Reference Letter on school letterhead stationery)

December 18, 20xx

[Recipient address- Line 1]
[Recipient address- Line 2]
[Recipient address- Line 3]
[Recipient address- Line 4]

To Whom It May Concern:

Re: Marion Melich – Practice Teaching Evaluation

I supervised Marion Melich's student teaching assignment at Wendell Miller Memorial High School during the fall term of 20xx. She also participated in an evening course I taught on "New Approaches to High School Teaching."

Marion did a solid job as a student teacher, particularly for her first time. She closely observed her fellow teachers and gradually developed her own approach to the teaching of English Literature. She went out of her way to give students individual attention and worked hard to make sure that all students understood the material. She learned to pace her lessons so that they kept the students' attention. She quickly developed a sense of confidence about her own teaching style but remained open to the suggestions of others.

Marion got along very well with her cooperating teacher. The students clearly liked and respected her and were sad to see her leave.

Marion is bright, conscientious, and personable. I believe she will make an excellent teacher if she decides that this is the career she would like to pursue.

Sincerely,

Hermine Grant
Superintendent, Student Teaching

Reference 19: Employment - Student Teacher (neutral)

(print teaching-related Reference Letter on school letterhead stationery)

April 30, 20xx

[Recipient address- Line 1]
[Recipient address- Line 2]
[Recipient address- Line 3]
[Recipient address- Line 4]

To Whom It May Concern:

<u>Student Teacher Reference – David Goss</u>

David Goss was a student teacher under my supervision for ten weeks from January 15, 20xx to April 12, 20xx at Oakmount Valley Secondary School. During that term David taught three Algebra 1 classes.

Oakmount is a comprehensive high school in an urban setting with a large and diversified student population. Overall, David adapted quickly to the "inner city" situation and progressed well as a student teacher. He demonstrated good command of the subject matter, was very cooperative, and a pleasure to work with. He was always receptive to my suggestions and was continuously striving to improve his teaching methods.

David did encounter a few discipline and control problems with two of his large classes. However, after focusing on improving his class management skills and techniques, David was able to better manage these classes. I believe that this is one area that David needs to work on. In spite of his class management problems David was able to develop good relationships with the majority of students. He was well-liked by his fellow teachers.

David is a great person and very dedicated to his teaching aspirations. I am sure that with his kind of commitment, coupled with continued experience and training, David will be a fine addition to the teaching profession.

Sincerely,

Ian Salter
Department Head, Mathematics

Reference 20: Employment - Explain Departure

(Corporate Reference Letter is printed on corporate letterhead stationery)

395 North Summit Street
Arkansas City, KS
67005

December 15, 20xx

To Whom It May Concern:

RE: Reason For Leaving – Susan Williams

The purpose of this is to explain why Susan Williams left our company after less than one year of employment.

Unfortunately, Susan is an innocent victim of corporate restructuring and downsizing. When we hired Susan as an online advertising specialist 11 months ago we had high hopes that the Internet-based economy would quickly pick up, and that Susan would be a part of our move to the next level in the industry.

Things did not turn around as quickly as we had hoped. So, in August our Board of Directors decided to impose a major restructuring and downsizing program on the online operations of the company, particularly in marketing and advertising. Susan was one of the people in the advertising department with the least amount of seniority, so we had to let her go.

This is to confirm that during the period Susan Williams worked with us we found her performance to be above average. In fact, the head of our advertising department tells me that he would re-hire Susan in a hurry if our fortunes turned around again.

If you require further information, I can be reached at (316) 442-0955.

Sincerely,

Margaret Amaroso
Director, Human Resources

SAMPLE TEMPLATES – COLLEGE-RELATED REFERENCE LETTERS

The following pages contain eight (8) real-life templates of reference letters written for various college-related situations.

Notes Re: Sample Letter Formats

- Although the sample templates in this guide are based on actual situations involving real people, identifying details have been altered to protect privacy.

- All of the sample templates in this guide have been reduced in size slightly to fit the book's page format which allows for page headers and footers.

- The samples use font size of 11 points rather than the 12 points I recommend as ideal. Also, the top/bottom and right/left margins have been adjusted to fit the margins of this book. You may want to readjust them for an actual letter.

Reference 1: College - Graduate Studies (Education)

(print Reference Letter on corporate letterhead paper)

September 5, 20xx

[Recipient address- Line 1]
[Recipient address- Line 2]
[Recipient address- Line 3]
[Recipient address- Line 4]

Dear Admissions Officer:

<u>**Letter of Reference - Jeanine Chambers**</u>

I am writing this letter of reference at the request of Jeanine Chambers who is applying for your graduate program in Educational Studies. She has asked me to provide this general letter that she may show to any prospective graduate school to which she may apply.

This letter is written on a confidential basis and therefore should be shared only with individuals in your school/program who are directly involved in the selection process. In addition, this letter should not be shown to Jeanine Chambers who has waived the right to see the letter.

I have known Jeanine Chambers as an undergraduate student for the past two years. During that time she was a student in three of my courses: Media, Technology and Computers and Education. These were both small third year classes so I was able to get a good look at Jeanine's performance and capabilities. Number one, she is an excellent communicator, both orally and in writing. Her papers and presentations were without exception, excellent. Her grade point average of 3.87 (4.0) for the two courses reflects her dedication to excellence. I was always impressed by her depth of research and she surprised me a number of times by taking an innovative and thought provoking look at generally accepted standards and norms. She wrote a wonderful paper on "Education and the Fear of Technology" which I suggest you read if you ever get the chance.

Jeanine was an active participant in class discussions and was convincing in making her points. If she has one Achilles heel it is her tendency to be somewhat impatient, bordering on intolerant, at times with the points of view proffered by some of her less capable peers. Although to be fair, she was often reacting to poor research and/or preparation on the part of the peer.

Based on my association with Jeanine Chambers, I believe that she would be a welcome addition to any graduate program in Pedagogical Studies, particularly if there is a strong research component.

Yours collegially,

Barbara Miller, Ph.D.
Department Head, Secondary Education

Reference 2: College - Undergraduate Scholarship

(print personal Reference Letter on regular stationery)

November 1, 20xx

Scholarships Office
Admissions, Recruitment and Registrar's Office
James Administration Building
845 Sherbrooke Street West
Montreal, Quebec
H3A 2T5

Dear Admissions Staff:

<u>**Letter of Reference - Samantha Brennan**</u>

I am writing this reference letter in support of Samantha Brennan's application for one of your entrance scholarships for first-year undergraduate students. As a neighbor, friend of the family, and Girl Guide leader I have known Samantha since she was seven years old.

I believe that Samantha would be an excellent candidate for one of the scholarships you are offering. In the eleven years that I have known Sam, she has been a wonderful role model for my two children. As a regular babysitter at my home over the past five years, she has had an important positive influence on both of my kids. Sam is the person who taught them to be gentle with their new puppy. She demonstrated unusual kindness, compassion, and understanding when their pet rabbit died unexpectedly. She also showed them how important their school work is in the scheme of things, and she was an excellent tutor when required.

Samantha regularly demonstrates empathy and compassion for those less fortunate. For her final two years of high school, twice a week after school she worked with a disabled child, teaching him how to swim. Last summer she was a volunteer coach of a neighborhood tee-ball team although she had to fit that in with her job as a camp counselor. She is well liked throughout our little community.

I believe that these attributes demonstrate that Samantha Brennan is of exceptional character. She is a person who will add character to your university and student body. I would ask that you please consider awarding her one of your scholarships to supplement her education.

Sincerely yours,

Wendy Montgomery

Reference 3: College - Teaching Award

(print award nomination Reference Letter on corporate letterhead stationery)

February 25, 20xx

Selection Committee
Washington University School of Medicine
1425 Lexington Avenue
Pullman, WA, 99164

Dear Committee Members:

Re: Washington U. Trustee Teaching Award - Nomination of Dr. A. Dharwan

I am writing this letter of reference in support of Dr. Anand Dharwan's nomination for a Washington University Trustee Teaching Award.

I have known Dr. Dharwan for the past 18 months, as a fellow in the division of Clinical Pharmacology/Gastroenterology and Hepatology. During that period I worked very closely with him in the clinic and on various research projects. As a result, I have become very familiar with Dr. Dharwan as a teacher. Accordingly, I am writing this because it is my sincere belief that he is eminently qualified to be a recipient of a Washington State University Trustee Teaching Award.

Over the past year and one-half Dr. Dharwan has demonstrated to me time and again that he is an exceptional communicator and gifted teacher. In my opinion, he is the perfect model for a teacher in a school of medicine. He is a highly dedicated professional who knows how to motivate his students to strive for excellence. Both in the lecture hall and in person, Dr. Dharwan is a dynamic and powerful communicator who possesses the gift of being able to make complex subjects understandable. In spite of these formidable gifts, he is a humble and approachable person who loves to share his extensive knowledge with others.

Dr. Dharwan has played a major role in helping me choose a career in Clinical Pharmacology Gastroenterology/Hepatology. Under his tutelage I have been motivated to strive for and achieve exceptionally high standards in patient care. In our work together on several research projects he provided me with outstanding mentorship in the disciplines of: study design, protocol writing, data analysis, and reporting. In many areas, both personally and professionally, Dr. Dharwan has taught and encouraged me to meet challenges that I had never before thought possible.

In closing, it is with sincere conviction that I enthusiastically recommend that Dr. Dharwan be selected as a recipient of a Washington University Trustee Teaching Award.

Sincerely,

Reiner G. Pinter, M.D.

Reference 4: College – Exchange Program

(print student nomination Reference Letter on corporate letterhead stationery)

April 15, 20xx

International Student Exchange Program
Participant Selection Committee
McGill University
Admissions and Registrar's Office
1735 Sherbrooke Street West,
Montreal, Quebec H3A 2T5

To Whom It May Concern:

<u>Letter of Reference – Lindsay Hamilton</u>

I am writing this reference letter at the request of Lindsay Hamilton in support of her application to participate in the International Student Exchange Program (ISEP) through McGill University. It is my understanding that this letter is being submitted on a sealed and confidential basis.

Lindsay Hamilton was one of my students in World History, 1914 to 1945 during the Fall Semester, 20xx. Since it was a relatively small class of about 20 students, I believe I was able to form a reasonably accurate picture of Lindsay's academic performance in the classroom and to assess of her capabilities.

Lindsay struck me as a serious, conscientious, and motivated student. She demonstrated this by attending all of her classes in a punctual manner and participating fully. She was also an active and articulate participant in class discussions. The presentation that Lindsay and her seminar group peers did on "the historian" was well-researched and ably presented, and provided an excellent basis for class discussion. Her academic performance over the four-month term warranted a solid mark of A minus.

Based on my four-month assessment of Lindsay, I view her as a bright, self-possessed and mature young woman with a lot of potential for academic growth. I understand that Lindsay has been an active international traveler since she was a child, and has extensive travel experience in parts of Asia and Europe. These experiences should serve her well in any type of international study or exchange program in which she participates.

In closing, based on my four-month exposure to her, I believe that Lindsay Hamilton would make an exemplary candidate for ISEP or other such programs.

Sincerely,

Michael Kirby, Ph.D.
Professor, Department of History

Reference 5: College – Business Undergrad Program

(print Reference Letter on company letterhead paper)

April 20, 20xx

Addressee Line 1
Addressee Line 2
Addressee Line 3
Addressee Line 4

Dear Sir/Madam:

Subject: Letter of Reference – Winston Prince

I have known Winston Prince since September 20xx when we met during the Orientation Programme as we began our careers at Bromberg L.P. As a colleague, I quickly developed a sense of respect and admiration for Winston. Even in those early days, it was evident to me that he was a bright, thoughtful, and articulate person. I could also see that he was very passionate about whatever he did or believed, with a natural ability to engage others.

Winston and I both started our careers in analytics. His thirst for knowledge meant that he was constantly learning about different areas of the markets. It was not long before we were all turning to him for help with client questions. Our careers followed similar paths into Sales, where Winston's passion and work ethic quickly earned him the recognition and respect of management. He was soon promoted to manage the North of England account.

Working with Winston on a number of projects and sales teams I could see that he was a highly focused, energetic and driven professional. He consistently took on challenging roles and then performed over and above what was expected of him. It was the norm for him to over-deliver.

Winston is a natural leader and doesn't hesitate to assume a leadership role. Due to frequent requests from his peers for help, he took it upon himself to organise weekly clinics on Excel and Fixed Income software, in which he answered his colleague's questions. With his quick intellect and natural sense of humour, Winston added a lot of charm and wit to his seminars and workshops. In fact, he is one of the rare few who can keep an audience interested in subjects as dry as "bond math swap valuation".

I always admired Winston's sense of purpose and discipline, when during our train journeys back to London he would find time to run through his CFA flash cards. He has an amazing ability to focus and compartmentalize, and he never loses sight of his goals.

I believe that Winston's departure was a great loss to our sales team and to the company as a whole.

Based on my experience working with Winston Prince, I can strongly recommend him as a candidate for any business program to which he might apply. I am confident he would make an invaluable contribution to that institution. If you have further questions please contact me at + 44 xx xxxx xxx xxx.

Sincerely,

Tom Mendes
Senior Marketing Associate
Bromberg L.P.

Reference 6: College – Med. Residency: Emergency

(print medical residency Reference Letter on corporate letterhead stationery)

February 25, 20xx

Dr. Raymond Lightstone
Director General
University of Kansas Medical Center
1845 Sunset Ridge, Kansas City, KS 66160

Dear Dr. Lightstone:

I am pleased to recommend Nadim Sharma for the residency program at your hospital to which I understand he recently applied. Nadim worked with me as an intern during his internal medicine rotations at Johns Hopkins, Medical Center from January to October this year.

I have been working as physician since 1980, and rarely have I seen an intern with so much enthusiasm and discipline as Nadim. From the very first day he impressed me in a number of ways. While he quickly mastered the art of history-taking and conducting thorough physical examinations, it was his ability to establish warm and empathetic relationships with patients that always caught my eye. He got actively involved in ward duties and routine diagnostic procedures and willingly stayed up late discussing cases and monitoring patients. In addition, he eagerly spent his free hours sitting in the wards, interacting with patients and educating them about their conditions.

While assisting me in the emergency room, Nadim gained first hand experience managing diverse cases including, but not limited to, shock, trauma, seizures and cardiovascular emergencies. I remember one occasion in particular when, near the end of a long ER night shift, a patient was brought in unconscious, in a state of severe shock. Even though Nadim was at the end of a wearying shift and I told him he was free to go home, he insisted on staying. His statement to me was simple and to the point, "Would I go home if this poor old man was my father?" His professional attitude and compassionate words left me overwhelmed.

More than anything, Nadim is an honest person who is not afraid to admit his ignorance. Once he asked me "What if I make some mistake, does that mean I am an irresponsible doctor?" He felt comforted when I assured him that every doctor will make mistakes at times, but every mistake provides an opportunity to learn if handled in the right way. Knowing him as well as I do, I am sure Nadim will not disappoint, and will probably exceed your expectations.

In summary, Nadim Sharma is an exceptionally bright and hardworking individual who throws himself enthusiastically into whatever he undertakes. Accordingly, I have no hesitation whatsoever in recommending him as a candidate for your residency program.

Sincerely,

Dr. Gordon Murphy

Reference 7: College – Med. Residency: Plastics

(print medical residency Reference Letter on corporate letterhead stationery)

October, 14, 20xx

[Recipient Address Line 1]
[Recipient Address Line 2]
[Recipient Address Line 3]
[Recipient Address Line 4]

To Whom It May Concern:

<u>**Letter of Reference – Kevin Stewart**</u>

I am very pleased to be able to write this Letter of Reference on behalf of Kevin Stewart in support of his application for a medical residency position in Plastic and Reconstructive Surgery.

I have known Kevin Stewart since July 20xx in my capacity as Supervisor, Plastic and Reconstructive Surgery at the Ontario General Hospital (OGH), a Queen's University teaching hospital in Kingston, Ontario. I supervised Kevin during both his Pre-Clerkship and his Clerkship rotations in Plastic Surgery at the OGH. I believe that puts me in an excellent position to make an accurate assessment of Todd's academic, scientific, and practical knowledge, skills and capabilities, as well as his overall personal qualities.

The following paragraphs summarize my assessments of Mr. Stewart's skills, abilities and attributes using the suggested "Guidelines to Referees" listed on the CARMS Website.

Cognitive Skills and Knowledge
Kevin is highly intelligent and very hard working. I found his level of knowledge to be beyond what I would normally have expected during a rotation assignment. He was highly motivated to learn and was a very quick study. His natural tendencies to be self-critical and always inquisitive enhanced his ability to learn and adapt. His knowledge and technical skills were noticeably superior, and he often performed at the level I would expect of a Resident.

Problem Solving and Patient Management
He displays excellent thought formulation and usually solves problems as they arise, quickly and effectively. His emergency room consults are always meticulous and easy for staff to follow. With patients and their families, Kevin is caring and sensitive and exhibits a combination of knowledge, confidence and empathy that quickly gains their attention and respect.

Behaviour and Attitudinal Skills
Kevin is truly a pleasure to work with and to be around. He is a highly motivated and hard working person with a very pleasant demeanor who embraces his work. Team members (i.e. attendings, residents, nurses, support), and patients alike, clearly enjoy being around a person with such a keen disposition. I can always count on Kevin to deliver what is expected of him since he is highly dependable.

Communication Skills and Working Relationships
Feedback from those working and interacting with Kevin is always positive. Everyone enjoys working with him. With patients he takes the time to carefully explain their situation (i.e. investigation, management, diagnosis, treatment plan) in a way that gains their confidence and trust. He manages to take the time and effort to do this while still handling an above-average caseload.

Ability to Work In A Team
Kevin works very effectively, both independently, and as a team member whenever required. His overall positive attitude, coupled with his excellent people skills, and his willingness to participate, make him an

ideal team member. Kevin is very friendly and likeable fellow; his great sense of humor and his self-deprecating manner quickly put people at ease.

Motivation and Punctuality
Kevin is an enthusiastic, highly-motivated, and driven medical student with a thirst to learn. As a result; he always presents on time, often takes on more than his share of the work, and stays late when warranted. Whenever required to attend to an emergency when on-call, he has responded exceptionally quickly and usually arrives within a few minutes of the call.

Sense of Responsibility
My experience working with Kevin has shown him to be a highly reliable and trustworthy person with a strong sense of responsibility. He never cuts corners; his consults are always thorough and precise, and are always trusted by those who use them. He is conscientious and disciplined in the way he approaches everything he does and he always acquits himself as a professional.

Procedural Skills
Kevin is very surgically-minded and demonstrates superior procedural and technical skills. He tends to always be one step ahead of what I would expect of a medial student in training. He makes extra effort to acquire new skills and applies them at every opportunity. He also has excellent clinical judgment and his clinical evaluations are always above average.

I believe that most of Kevin's superior skills, abilities and attributes can be summarized by describing one particular case he attended to while under my supervision. It involved a 50-year old woman who presented with extensive tissue evulsion injuries after a serious car accident. Kevin responded quickly to the call in the middle of the night, and within minutes he promptly attended to the case in a highly effective manner. He correctly assessed the situation, ordered the appropriate investigations, and then initiated appropriate treatment including proper sedation and local anesthesia, as well as performing a combination of primary closure and full thickness skin grafting involving more than 300 stitches.This was an exceptional performance for a medical student.

In closing, my experience working with Kevin Stewart has shown him to be a truly exceptional medical student. He has clearly demonstrated to me that he has the potential to become a highly competent practitioner of Plastic and Reconstructive Surgery. I therefore recommend him without hesitation or qualification for a postgraduate residency position in that specialty.

Please don't hesitate to contact me should you have any questions or require additional information.

Sincerely,

Dr. R. Marinaro

Reference 8: College – Med. Residency: Neurosurgery

(print medical residency Reference Letter on corporate letterhead stationery)

October, 18, 20xx

[Recipient Address Line 1]
[Recipient Address Line 2]
[Recipient Address Line 3]
[Recipient Address Line 4]

To Whom It May Concern:

<u>**Letter of Reference – Todd Bryson**</u>

I am pleased to be able to write this Letter of Reference on behalf of Todd Bryson in support of his application for a medical residency position in Neurosurgery.

I have known Todd Bryson since May 20xx in my capacity as Chief of Neurosurgery at the Regional General Hospital (RGH), a McMann University teaching hospital in Montreal. I supervised Todd during both his Pre-Clerkship and Clerkship surgery rotations in at the RGH. I believe that puts me in an excellent position to make an accurate assessment of Todd's academic, scientific, and practical knowledge, skills and capabilities, as well as his overall personal qualities.

Todd is highly intelligent and very hard working. I found his level of knowledge to be beyond what I would normally have expected during a rotation assignment. He was highly motivated to learn and was a very quick study. His natural tendencies to be self-critical and always inquisitive enhanced his ability to learn and adapt. His knowledge and technical skills were noticeably superior, and he often performed at the level I would expect of a junior resident.

Mr. Bryson displays excellent thought formulation and usually solves problems as they arise quickly and effectively. His emergency room and off-service consults are always meticulous and easy for staff to follow. With patients and their families, Todd is caring and sensitive and exhibits a combination of knowledge, confidence and empathy that quickly gains their attention and respect.

Todd is also a pleasure to work with and to be around. He is a highly motivated and hard working person with a very pleasant demeanor who embraces his work. Team members (i.e. attendings, residents, nurses, support), and patients alike, clearly enjoy being around a person with such a keen disposition. I can always count on Todd to deliver what is expected of him since he is highly dependable.

Feedback from those working and interacting with Todd is always positive. Everyone enjoys working with him. With patients he takes the time to carefully explain their situation (i.e. investigation, management, diagnosis, treatment plan) in a way that gains their confidence and trust. He manages to take the time and effort to do this while still handling an above-average caseload.

Todd works very effectively, both independently, and as a team member whenever required. His overall positive attitude, coupled with his excellent people skills, and his willingness to participate, make him an ideal team member. Todd is very friendly and likeable fellow; his great sense of humor and his self-deprecating manner quickly put people at ease.

I view Todd as an enthusiastic, highly-motivated, and driven medical student with a thirst to learn. For example, he always presents on time, often takes on more than his share of the work, and stays late when warranted. Whenever required to attend to an emergency when on-call or when asked to cover for his colleagues, he has responded exceptionally quickly and always arrives within a few minutes of the call.

My experience working with Todd has shown him to be a highly reliable and trustworthy person with a strong sense of responsibility. He never cuts corners; his consults are always thorough and very precise, and are always trusted by those who use them. He is conscientious and disciplined in the way he approaches everything he does and he always acquits himself as a professional.

Todd is very surgically-minded and demonstrates superior procedural and technical skills. He tends to always be one step ahead of what I would expect of a medical student. He makes extra effort to acquire new skills and applies them at every opportunity. He also has excellent clinical judgment and his clinical evaluations are always above average. Knowing that Todd has a special interest in Neurosurgery I tried to involve him whenever possible in various aspects of my peripheral neurosurgery practice (e.g. carpal tunnel clinic, digit retransplantation clinic) to give him a basic grounding in those techniques.

An interesting point about Todd's background is his demonstrated ability to obtain research funding. Through my discussions with him I learned about his work conducting basic neuroscience research and how he was able to successfully attract research grants to fund that work through such programs as NSERC, OGS, and RESTRACOMP.

In closing, my experience working with Todd Bryson has shown him to be an exceptional medical student. He has clearly demonstrated to me that he has the potential to become a highly competent practitioner of Neurosurgery. I therefore recommend him without hesitation or qualification for a postgraduate residency position in that specialty.

Please do not hesitate to contact me should you have any questions or require additional information.

Sincerely,

Dr. W. Blum

SAMPLE TEMPLATES – CHARACTER-RELATED REFERENCE LETTERS

The following pages contain seven (7) real-life templates of character-related reference letters written for various situations.

Notes Re: Sample Letter Formats

- Although the sample templates in this guide are based on actual situations involving real people, identifying details have been altered to protect privacy.

- All of the sample templates in this guide have been reduced in size slightly to fit the book's page format which allows for page headers and footers.

- The samples use font size of 11 points rather than the 12 points I recommend as ideal. Also, the top/bottom and right/left margins have been adjusted to fit the margins of this book. You may want to readjust them for an actual letter.

Reference 1: Character - Friend

(Character Reference Letter can be printed on company letterhead or standard paper)

501 Kemper Ave.
St. Louis, MO, 63139

October 14, 20xx

To Whom It May Concern:

RE: Character Reference – Jason Sunderland

The purpose of this is to provide a character reference for Mr. Jason Sunderland who I have known as a classmate, roommate, and friend for a period of five years.

I first met Jason in our freshman year at Adirondack College. We were both studying a general arts program there and became acquainted through a number of common classes that we shared. By second year, we had become friends and decided to take an off-campus apartment together. We shared that living arrangement until we both graduated last year.

Accordingly, having gotten to know Jason so well over the past few years, I believe puts me in a position to provide you with a pretty accurate assessment of his character.

As a student, Jason was a hard-working and highly committed to his education. I believe that his excellent transcripts will attest to that fact. In addition, he was quite involved in a number of extra-curricular activities including the track and field team and the school newspaper. In fact, in his last two years he was Assistant Editor of the "Campus Inquirer." Outgoing, and always willing to help someone out, Jason was very popular with his fellow students.

As a roommate, Jason was a great choice. He was very neat and tidy at all times and he liked things in the apartment to be kept orderly. He made a point of cleaning his own room and the common living areas on a regular basis. He socialized occasionally at home but was always respectful of my needs, and he and his guests kept the noise down and ended their activities at a reasonable hour.

As a friend, Jason Sunderland is a standout. He is a loyal, honest, considerate, and supportive individual who has the ability to see and understand things from another person's perspective.
He is a great direct communicator and knows how to raise and discuss common living issues and problems in a non-threatening manner. He is hyper-sensitive and is always tuned into how another person might "feel" in a given situation. He likes to have fun too. During our years at school we maintained an ongoing friendly rivalry on the squash courts.

To tell the truth, I really can't think of anything of consequence on the negative side of the personality ledger when it comes to Jason. All in all, I would have to say that Jason Sunderland is a fine, well-balanced person with an abundance of positive qualities.

Sincerely,

Ronald Marrion

Reference 2: Character - Colleague

(Character Reference Letter can be printed on company letterhead or standard paper)

Guildhall
PO Box 270
London EC2P 2EJ.

October 14, 20xx

[Recipient address- Line 1]
[Recipient address- Line 2]
[Recipient address- Line 3]

Dear Sir/Madam,

Reference: Randall Morton

I have known Randall Morton since September 20xx when we met as colleagues during the Orientation Programme as we began our careers at Advanced Applications LLC.

As a peer, I quickly developed a sense of respect and admiration for Randall. Even in those early days, it was evident to me that he was a bright, thoughtful, and articulate person. I could also see that he was very passionate, with a natural ability to engage others.

Randall and I both started our careers in analytics. His thirst for knowledge meant that he was constantly learning about different areas of the markets. It was not long before we were all turning to him for help with client questions. Our careers followed similar paths into the field of sales, where Randall's passion and work ethic quickly earned him the recognition and respect of management. He was soon promoted to manage the Northern Ireland account.

Working with Randall on a number of projects and sales teams I could see that he was a highly focused, energetic and driven professional. He consistently took on challenging roles and then performed over and above what was expected of him. It was the norm for him to over-deliver.

Randall is a natural leader and doesn't hesitate to assume a leadership role. Due to frequent requests from his peers for help, he took it upon himself to organize weekly clinics on Excel and Fixed Income software, in which he answered his colleague's questions. With his quick intellect and natural sense of humor, Randall brought a lot of charm and wit to his seminars and workshops. In fact, he is one of the rare few who can keep an audience interested in a subject as dry as "bond market swap valuation".

As a person, Randall is an outstanding individual. I always admired his sense of purpose and discipline, when on our train journeys back to London he would find time to run through his CFA flash cards. He has an amazing ability to focus and never loses sight of his goals. At the same time, he is able to balance that sense of focus and remain available as a person to friends and colleagues. In fact, Randall was so well liked here that many on our sales team experienced a serious sense of loss when he left the company.

In summary, based on my experience as both a colleague and friend of Randall Morton, I can strongly recommend him as a person and as a candidate for your business program. I am confident he would make an invaluable contribution to your institution. If you have any further questions please don't hesitate to contact me at + 44 20 5785 453 790.

Sincerely.

Raymond Lancer
Senior Market Analyst

Reference 3: Character – Domestic Services

(Reference Letter from a private citizen is printed on regular stationery)

9528 Carling Ave.
Apt. 1508
Ottawa, Ontario
K2B 8M5

January 25, 20xx

[Recipient address- Line 1]
[Recipient address- Line 2]
[Recipient address- Line 3]
[Recipient address- Line 4]

To Whom It May Concern:

RE: Character Reference - Manuela Gonzalez

I have known Manuela Gonzalez since January 20xx when she began living-in with my ex-wife and caring for my daughter Charlotte.

Due to a divorce, I don't live in the same house as Manuela and my daughter, but I believe that over the past 20 months or so, I have had enough contact with her, both directly and indirectly (through my daughter's behavior), to be able to possess a reasonably accurate impression of her attributes and character.

Accordingly, based on all my personal dealings with Manuela (as well as comments made by Charlotte's mother), I find her to be a very kind, loving, and caring individual. I am very comforted to have such a person responsible for the day-to-day care of my daughter. In addition, I have also noted that Manuela is hard-working, responsible, conscientious, and honest. She has displayed good judgment, tactfulness, diplomacy, and co-operation in all of the contacts I have had with her.

Therefore, based on my knowledge of her, I would not hesitate to state that Manuela Gonzalez is a very dependable individual of seemingly impeccable character.

If you require further information, please don't hesitate to contact the undersigned at (819) 997-4295 or (613) 232-4158.

Sincerely,

Peter R. Shannon

Reference 4: Character – Long-Time Client

(print business-related Reference Letter on corporate letterhead stationery)

December 12, 20xx

[Recipient address- Line 1]
[Recipient address- Line 2]
[Recipient address- Line 3]
[Recipient address- Line 4]

To Whom It May Concern:

Re: Character Reference - Olivia Thyani

The purpose of this is to provide a character reference for Mrs. Olivia Thyani whom I have known as a client and friend for 19 years.

I first met Olivia early in my career as an Investment Advisor while she was an Economics Professor at York University and President of Sidbex Ltd. Since that time we have enjoyed a professional working relationship as Advisor and Client. I believe this puts me in an ideal position to provide you with an accurate assessment of Olivia's character.

As a client, Olivia has always conducted herself in a very professional manner. Her account has consistently been in good standing, and we greatly value her business with our organization. Olivia has demonstrated excellent ethics and integrity in all of her business dealings with us. I have also been very impressed with her knowledge and understanding of all financial matters and have, on occasion, appreciated her advice.

As a friend, I can tell you that Olivia is a very caring, understanding and supportive person who values her relationship with her two children and her three grandsons. She makes a special effort and takes a great interest in encouraging her grandsons as they embark on their educational experiences. She is also an active and accomplished hiker and skier.

In summary, it has been my great pleasure to have known Olivia over the past 19 years, and I have the utmost respect for her as both a professional and a friend.

If you would like any further information regarding my knowledge of Olivia, you are welcome to contact me at (514) 734-5345.

Sincerely,

Nasser Ismali
Senior Investment Counselor

Reference 5: Character - Community Colleague

(print Reference Letter on corporate letterhead paper)

April 29, 20xx

[Recipient address- Line 1]
[Recipient address- Line 2]
[Recipient address- Line 3]
[Recipient address- Line 4]

To Whom It May Concern:

<u>Reference – Charles Renfroe</u>

I have known Charles Renfroe since 1999. Charles and I belong to the same church where we have served together on a number of committees involved in community service in the greater Hartford area.

In the four years I have known Charles I have been impressed with his dedication to any endeavor he has been involved with. He consistently demonstrates a giving nature and is very generous with his free time. For example, last year he participated in a mission trip with his young adults group to build housing for deprived people in Guatemala. To be part of that project he had to finance his own trip and he also lost one week of wages. Two years ago he used his own summer vacation to lead one of our youth groups on a wilderness canoe/camping trip.

Charles demonstrated his honesty and integrity just recently when he found a purse in a parking garage that contained over $300 and many credit cards. He not only turned the purse in at the hotel front desk, but he looked up the phone number of the owner and called her to make sure she was aware of where to locate it. Due to his accounting training he also serves as the trusted treasurer of two service clubs.

I trust that you will give serious consideration to selecting Charles for the position for which he is being considered. I believe that he would be an excellent addition to any organization.

Yours in service,

Reggie Idlewild

Reference 6: Character – Committee Member

(print a Reference Letterr from a private citizen on regular stationery)

532 Drummond St.
Perth, ON
K7H 4N5

November 20, 20xx

Ms. Jean Rogers
Chairperson
Adult Programs - Community
 Consultation and Advisory Committee
War Memorial Hospital
Perth, ON, K7N 4R2

<u>CONFIDENTIAL</u>

Dear Jean Rogers:

This is in reply to your recent letter to me in which you asked my opinion of Mr. Robert Jackman as a possible member of your Advisory Committee.

Based on your letter, it is my clear understanding that you have contacted me on this in complete confidence, because I previously held the position of Chairperson of your committee. On that strictly confidential basis, I have the following comments to make about Mr. Jackman as they relate to him as a possible candidate for your committee.

I have known Robert Jackman for more than 25 years, nine of which we worked together as insurance underwriters for the Mutual-Benefit Insurance Group. During that 25-year period Mr. Carson and I also served on a number of local community committees together.

Although I respect Mr. Jackman as a highly successful insurance underwriter, I must say in all honesty that I cannot recommend him for service on your committee. Without resorting to personal assassination, I have to state that Mr. Jackman can be very difficult to work with at times, especially in a committee situation. He seems to have a strong need to always be in charge, whether he is chairperson or not. He often tends to be autocratic and obstinate, and is almost invariably in disagreement with the vast majority of his colleagues.

You may recall two years ago when the Parks and Monuments Committee resigned en masse. That unfortunate group resignation was a direct result of Mr. Jackman, who was chairperson at the time, unilaterally making unpopular decisions without consulting with his committee members. This is just one example of the kind of problems Mr. Jackman has caused over the years. There are numerous others, which I won't go into here.

In closing, based on my experience working with Robert Jackman, both professionally and in community service over a 25-year period, I cannot in good conscience recommend him to you for appointment to your advisory committee.

If you have any questions on this, please don't hesitate to call me at 371-5289.

Sincerely,

Randolf Smithfield

Reference 7: Character - Rehabilitated Ex-Convict

(print Ex-Convict Character Reference Letter on personal stationery)

95 Quackenbush Court
Albany, NY 12207

June 12, 20xx

Addressee Line 1
Addressee Line 2
Addressee Line 3
Addressee Line 4

To Whom It May Concern:

Re: Reference Letter – Jacob Flynn

I am writing this letter on behalf of Jacob Flynn in the hope that it will help you understand how much he has changed and how much he has to offer as a person, in spite of certain transgressions in his past.

I have known Jacob for more than a decade, since we were classmates and teammates in high school. I remember him during that period as a kind and considerate person; one who I was proud to call my friend. He was a good student and very well liked by students and teachers, alike.

Sadly, during the summer after our graduation, Jacob's father and younger brother were both killed in a tragic car accident. That's when I began to see another side of Jacob I had never seen before. Somewhat understandably, he became deeply depressed and despondent over the loss of his family members and began to drink alcohol excessively. For about three years after that Jacob continued along that same self-destructive track, during which he essentially cut himself off from family and friends.

From what I observed, Jacob's conviction for impaired driving causing bodily harm two years ago was a sudden wake-up call for him which snapped him back into reality. I have spent a lot of time with Jacob since he was released from detention last summer and I am convinced that he sincerely regrets what he did and is ready to move forward with his life. I have witnessed him express deep remorse on more than one occasion and he has made serious efforts to make amends to those he injured. He is back at college, studying business management, after which he plans to open a small business. He attends meetings of Alcoholics Anonymous on a regular basis, and he works as a volunteer in two community outreach organizations.

In my opinion, Jacob Flynn has been fully rehabilitated and is already proving himself to be a productive member of society. I sincerely hope that you will see Jacob Flynn for what he has become and not what he was. He has proven that he deserves a second chance.

Sincerely,

Randall Whittaker

SAMPLE TEMPLATES – GENERAL REFERENCE LETTERS

The following pages contain six (6) real-life templates of reference letters written for various general situations.

Notes Re: Sample Letter Formats

- Although the sample templates in this guide are based on actual situations involving real people, identifying details have been altered to protect privacy.

- All of the sample templates in this guide have been reduced in size slightly to fit the book's page format which allows for page headers and footers.

- The samples use font size of 11 points rather than the 12 points I recommend as ideal. Also, the top/bottom and right/left margins have been adjusted to fit the margins of this book. You may want to readjust them for an actual letter.

Reference 1: General – Business

(Corporate Reference Letter is printed on corporate letterhead stationery)

3005 19th Street, NE
Calgary, AB
T2E 6Y9

November 20, 20xx

[Recipient address- Line 1]
[Recipient address- Line 2]
[Recipient address- Line 3]
[Recipient address- Line 4]

To Whom It May Concern:

RE: Customer Reference – Thompson Graphics Inc.

I have been asked to write this letter of reference because our company will no longer be operating its printing plant that has served Thompson Graphics Inc. for more than a decade.

Thompson Graphics has been one of our top customers for the past 15 years. Accordingly, I have no hesitation in recommending them as a company to do business with.

In addition to doing business with his company for many years, Ray Thompson and I go back to our university days over 25 years ago. So, I can also vouch for him as a great individual and a concerned and active citizen in this community.

As far as a company to do business with, Thompson Graphics is one of the best that we have ever dealt with. Its practice was always to pay our printing invoices within the 30-day time limit. We did significant amounts of business, especially during the past 5 years, and I cannot recall a late-payment situation involving that company. Billing disputes were rare, and those only required some minor additional documentation for clarification and resolution.

Thompson was one of the best companies that I have ever dealt with from a change-order and work scheduling perspective. We maintained a close communication with the company's production people and they always kept us apprised of their upcoming workload, so that scheduling jobs on our presses was never a problem. In addition, Thompson's graphics people always provided us with high quality finished artwork, and it was unusual for additional changes to be made after the plates had been produced.

Based on our experience, any printing company should be very pleased to be the one that Thompson Graphics chooses to do business with once we have closed our doors.

Sincerely,

Stewart Johannsen
President and CEO

Reference 2: General - Counseling Service

(print personal Reference Letter on regular personal stationery)

August 23, 20xx

[Recipient address- Line 1]
[Recipient address- Line 2]
[Recipient address- Line 3]
[Recipient address- Line 4]

To Whom It May Concern:

Letter of Reference – Megan O'Malley

I am writing this letter of reference to state how grateful my family is that Megan O'Malley came into our lives and the life of our Mother.

For some time, my siblings and I had been wondering just what we were going to do with my mother who was suffering increasingly from short term memory loss. Since none of us lived near her, we were mostly sitting on the sidelines watching her deteriorate based largely on sketchy reports from our father. Since mom was still driving, getting her hair done, and going to church we were all in a certain amount of denial about the exact state of her mental health. However, all of my suspicions were confirmed when I spent a three-week vacation with my folks last summer. I saw that mom's mental health had deteriorated even more than we had thought.

That's when I was fortunate enough to find Megan O'Malley through the local Community Connections Network. Megan quickly became my lifeline to understanding my mother's illness and for finding support resources to help care for her. She provided the family with detailed information on the stages of Alzheimer's. She also gave us valuable advice and put me in touch with the Alzheimer's Association in my hometown as well as where my folks live. Through Megan I was able to find a qualified nurse/case manager who guided me through the process of getting a neurologist to do an assessment of mom.

Megan also helped a great deal with the family's understanding of my mom's condition. She got us talking about mom's future care and urged us to lok at options such as assisted living programs. She briefed us on safety and how to handle day-to-day life with memory impaired people. She also spent time educating the primary caregiver - my dad - all about Alzheimer's and explained how people suffering from memory problems still want to maintain their dignity and have social contact. In addition, I was able to call Megan for individual counseling/consulting on many things including words and terms to use to communicate with my siblings, doctors and my mother.

Megan is a sensitive, wise, energetic and extremely compassionate family counselor. I highly recommend the services of Megan O'Malley if you and./or your family need a professional counselor/advisor to guide you through a family crisis due to illness or injury. Her fees are very reasonable and the advice and assistance you receive will be worth every penny.

Yours sincerely,

Patricia Hazelton

Reference 3: General - Bank Reference

(print bank Reference Letter on bank letterhead stationery)

November 20, 20xx

[Recipient address- Line 1]
[Recipient address- Line 2]
[Recipient address- Line 3]
[Recipient address- Line 4]

To Whom It May Concern:

<u>**Customer Reference – Richard K. Kelly**</u>

This letter will confirm that, as of this date, Richard K. Kelly of 1501-2400 Sherbrooke Street West, Montreal, QC, Canada H3C 2L3, is a customer in good standing with this Bank and has a history of meeting his business and personal banking commitments. Based on our experience with him, Mr. Kelly is a trustworthy and dependable person with whom to do business.

Our records state that Richard K. Kelly was born on July 14, 1952 in Toronto, Ontario, Canada..

Mr. Kelly has been a client at this branch for the past twelve (12) years. Over this period he has operated a full-service business account with an ongoing small business line-of-credit.

If you require further information, please contact the undersigned at (514) 862-2598.

Sincerely,

Brian Chadwick
Assistant Branch Manager

Reference 4: General - Customer Credit Reference

(print Credit Reference Lette on company letterhead paper)

August 7, 20xx

[Recipient address- Line 1]
[Recipient address- Line 2]
[Recipient address- Line 3]
[Recipient address- Line 4]

To Whom It May Concern:

Re: Credit Reference - Mr. Rico Liberatti

This is to confirm that Mr. Rico Liberatti has been a regular customer of good standing with Home & Country Hardware since 1993.

During that 10-year period Mr. Liberatti was one of our credit card holders. He made purchases from our stores on a regular basis and has an impeccable record of making regular payments on time. In fact, according to our credit database records, during the entire period that Mr. Liberatti was our regular customer he was never once late in making a payment. This is exceptional in the retail credit business.

Accordingly, based on our experience with Mr. Liberatti as one of our most valued credit card customers for over 10 years, I am pleased to recommend him as a highly dependable and extremely low credit risk.

If you require further information, please feel free to call me at (416) 232-9775.

Sincerely,

Jennifer Hale
Retail Credit Manager

Reference 5: General - Landlord Re: Pet Owner

(print landlord Reference Letter on landlord's business letterhead stationery)

July 20, 20xx

[Recipient address- Line 1]
[Recipient address- Line 2]
[Recipient address- Line 3]
[Recipient address- Line 4]

To Whom It May Concern:

I am pleased to be writing this letter on behalf of Eileen Atkinson and her cat Paddy. Eileen and Paddy have lived together in my building for the past seven years without incident. Eileen is an exemplary pet owner and Paddy was under her care and control at all times. In fact, it was rare that I ever saw Paddy, since he is essentially a true house cat.

When I inspected Eileen's apartment earlier this week there was no sign that anyone had even lived there with a pet. In fact, the unit was cleaner when Eileen moved out than it was when she took the apartment in 1996.

I therefore unconditionally recommend Eileen Atkinson and her cat Paddy as great tenants who should be welcomed with open arms into any building for which they apply.

Yours truly,

Roger Hastings
Building Superintendent

Reference 6: General - Tenant

(print tenant Reference Letter on landlord's business letterhead stationery)

August 10, 20xx

[Recipient address- Line 1]
[Recipient address- Line 2]
[Recipient address- Line 3]
[Recipient address- Line 4]

To Whom It May Concern:

<u>**Reference – Raleigh Radcliffe**</u>

This is to confirm that Mr. Raleigh Radcliffe was a tenant in excellent standing in the Active Life Condominium Complex for an eight-year period, from July 19xx to June 20xx.

I was the General Manager of the Active Life Condo Complex during the last six years of Mr. Radcliffe's tenancy. Throughout that period I always regarded him as a model tenant, and I often wished that more of our tenants could be like him.

Although he was a renter and not an owner, it was hard to tell by the way he maintained his yard and involved himself in the life of the community. He was very active on the Recreation Committee and was a long-time member of the Neighborhood Watch Patrol.

Overall, I would describe Mr. Radcliffe as dependable and involved. He always paid his rent and club fees on time. He frequently added value to the community through his own unsolicited actions. For example, he developed a new court booking system for the squash courts which allowed members to check availability and book court times over the Internet. He also was instrumental in obtaining City approval for the complex's garden plot project in 20xx.

Based on my association with him over a six year period as described above, I highly recommend Mr. Raleigh as an exemplary tenant who can be relied upon to meet his obligations and to add value to whichever community with which he becomes associated.

Sincerely,

Angelo Ricci
General Manager

COMMENDATION LETTERS

Commendation letters are unsolicited letters, which typically commend an employee to their supervisor or organization for something outstanding or noteworthy that the employee has done.

Normally, the employee would have to do something "above and beyond" what is routinely expected of them in their job to warrant such a letter.

Typically, these letters are written by co-workers, or managers from another area of the organization who were suitably impressed while supervising the person on a short-term project.

Commendation letters are also often written in the community service sector when citizens or organizations believe that an individual has made an exceptional contribution in serving the public as a volunteer.

DRAFTING TIPS – COMMENDATION LETTERS

For general letter writing tips, please refer back to the section earlier in this guide titled Letter Writing Guidelines on page 23.

The main thing that differentiates commendation letters from other types of recommendation letters is that they are almost always spontaneous and unsolicited, and normally come as a pleasant surprise to both the subject of the letter and its recipient.

Although there can be exceptions to this. For example; take the case when someone is asked to write a letter of commendation so that the subject can be nominated for a formal award.

Following are a number of tips to keep in mind when drafting a commendation letter for someone:

Think About It Carefully

Before writing an unsolicited commendation letter make sure in your own mind that it is truly warranted. This is not to discourage writing one of course, but they are meant to recognize a truly exceptional act or situation, well above the normal call of duty. Just think about it carefully and make sure it is the right thing to do under the circumstances. Once one is written and sent, you can't take it back.

Make sure that you are in a safe position to write such a letter. For example, if you are relatively new to the organization, you might want to wait a bit and do some research before you write such a letter. There could be lots of historical factors at play, some of which you may be unaware.

Do Some Basic Checking

In an organizational setting, before sending a commendation letter about someone it would be a good idea to get a hold of that person's job description to make sure that what you think is exceptional is a not simply a mandatory duty in their job description.

In fact, I would suggest once you have done your basic checking and are convinced that you want to go ahead, before you do, you might want to run the idea past the person's supervisor just to make sure you aren't stepping on someone's toes or are acting on incomplete information.

If it is truly a situation that warrants a commendation letter, the subject's boss will normally be pleased about the positive kudos that will be going to the employee. (After all, it's also an indirect compliment to the subject's boss).

Get Proper Background Information

If the commendation letter you are writing involves an employee in an organization, you need to make sure you get the details right. After all, your letter will be passed around senior management in the organization and it will be placed on the employee's personnel file, permanently. In addition, the employee will more than likely use it as reference material when applying for promotions or other jobs in the future.

So, make sure you address it to the appropriate person and title. That is usually the employee's direct supervisor. Check that the details such as position titles and the spelling of names are correct.

Brace Yourself For Repercussions

This may be a bit sad, but in a typical medium to large organization, it is often a reality. No matter how well-intentioned you may be in writing your commendation letter, there will be those who are angry, upset, or resentful.

Yes, that's right. In my 25+ years working in a variety of organizations, both large and small, I learned that they are great incubators for the polishing and festering of the human ego.

The vast majority of people who become aware of your commendation letter will be happy for the recipient and wish them well. Nevertheless, there will also be those who resent it. Some people may regard you as suspect for writing such a letter, wondering what your hidden agenda is. Others will be acting out their jealousy wondering why they've never received such a commendation themselves.

Even if you're a manager, you could still be resented by other managers. They might not like the fact that you've taken the time to write such a letter for your employee, and feel that it makes them look bad in front of their own employees.

But don't let any of this stop you. If you think it is the right thing to do, go ahead and write that commendation letter.

SAMPLE TEMPLATES - COMMENDATION LETTERS

The following pages contain a number of commendation letter templates written for various real-life situations.

Notes Re: Sample Letter Formats

- Although the sample templates in this guide are based on actual situations involving real people, identifying details have been altered to protect privacy.

- All of the sample templates in this guide have been reduced in size slightly to fit the book's page format which allows for page headers and footers.

- The samples use font size of 11 points rather than the 12 points I recommend as ideal. Also, the top/bottom and right/left margins have been adjusted to fit the margins of this book. You may want to readjust them for an actual letter.

Commendation 1: Corporate - Employee

(Corporate Commendation normally printed on corporate memo letterhead stationery)

MEMORANDUM

Date: November 31, 20xx

From: Monica Bates

To: Pablo Cassavetes
           Director, Research Support Programs

Subject: <u>Commendation – Hubert McConnell – Freight Transport Demand Project</u>

The purpose of this is to officially commend Hubert McConnell for his exceptional contribution throughout his assignment to the Freight Transport Demand Project (FTDP).

As you know, Hubert has been working on special assignment with the FTDP team for the past eight months. Now that he is about to return to your part of the organization I wanted to make sure that he gets some recognition for his significant and exceptional contributions to the project.

As a junior econometrician, Hubert's role in the project was pivotal to its timely and successful completion. It was Hubert who worked long hours, numerous nights and weekends with his small team of researchers, first specifying, and then testing the thousands of equations that had to be run. The quality of Hubert's written work was also exceptional. His regression analysis summaries were always very well written and rarely required revision.

As a colleague and project team member, Hubert was also outstanding. His upbeat enthusiasm for the project was infectious, and he seemed to motivate the entire project team. He was very well –liked by all team members, and in effect he became "unofficial" deputy project manager.

In closing, I would like to say that I have worked with many junior economists and econometricians over the years and have never run across one who was as professional and productive as Hubert McConnell was on the FTDP. I believe that the organization as a whole should recognize his exceptional contribution to a major project.

Please let me know if you have any questions or comments.

Catherine Potvin
Director, Econometric Research

cc: Hubert McConnell
          Personnel file – H. McConnell

Commendation 2: Customer Service - Employee

(Commendation from a private citizen normally printed on standard stationery)

4590 Green Avenue
Montreal, QC

September 20, 20xx

Ms. Vivien Wong
Manager, Guest Services
Mount Royal Hotel
4900 Sherbrooke Street
Montreal, H5S 3T7

Dear Ms. Wong:

<u>**RE: Exceptional Service – Mr. Eduardo Perez**</u>

My name is Paul Kubecki. I have been a member of your hotel's health and fitness club for over five years. I am writing this to you in your capacity as manager responsible for the club.

The sole purpose of this letter is to draw your attention to the exceptional level of service that has been provided by Mr. Eduardo Perez since he joined the staff of the health club, some 18 months ago. I normally wouldn't take the time to write a letter like this, but in Eduardo's case, I just had to because he has made such a difference to the level of service that members now receive.

From the day he joined the team at the club it was clear that Eduardo was different from those who had preceded him. It was obvious from the beginning that he has a clear understanding of what customer service is all about, and he knows how to deliver it to club members.

Among many other things, Eduardo is extremely courteous, thoughtful, and kind in all of his dealings with members. In addition, he is very effective in running club matters. For example, previously, when a machine would break down (i.e. a treadmill) it would take anywhere from 1 to 2 weeks for it to be repaired due to lack of follow-up by health club staff. Now, when a machine breaks down, Eduardo makes it his personal mission to see that things are followed-up. Since he arrived, repairs are always done within 24 to 48 hours.

Before Eduardo, it was a common occurrence for the water coolers to be left unfilled and the tissue boxes that should be kept outside of the squash courts would never be replaced without members complaining. Now with Eduardo in charge, these small but annoying things just don't happen anymore. He has instituted a "walk around" system whereby members of the health club staff must check all facilities at the beginning of their shift to see if anything needs attention.
This has made a big difference and has definitely been noticed by members.

I have discussed my intention to write this letter with a number of regular members and they enthusiastically supported the idea that Eduardo be acknowledged for his exceptional efforts.
We trust that Eduardo Perez will somehow be recognized for delivering a superior level of customer service to members of the Mount Royal Hotel, Health and Fitness Club.

If you would like further details, please don't hesitate to contact me at (514) 989-7299.

Sincerely,

Paul Kubecki

Commendation 3: Teacher - From Parent

(Commendation from a private citizen normally printed on standard stationery)

45 Muskoka Drive West
Orillia, ON
L3V 7T5

August 28, 20xx

Ms. Patricia Morton
Couchiching District Secondary School
78 Andrew St. S.
Orillia, ON
L4S 3R2

Dear Ms. Morton:

The purpose of this is to thank you for the positive influence you have had on our daughter Samantha this past school year.

Up until this year, Samantha had problems every year with her English language studies. This year we have noticed a dramatic turnaround, and it is clear to us that it is directly attributable to your teaching methods. Samantha's average in English Language Studies is now in the mid 80s when in previous years, she has never received an average over 65% in those subjects.

It also shows in Samantha's general attitude towards all of her subjects, and school in general. For the first time that we can remember, she looks forward to going to school and to your English Literature and Drama classes in particular. Her overall average for all subjects has increased this year by over 10 points, to above 80%.

Veronica and I feel that the changes in Samantha's attitude and marks have been so dramatic that we wanted to formally thank you for your efforts on our daughter's behalf. We have also taken the liberty of sending a copy of this letter to your Principal, Jackson Davies.

Thank you so much for what you have done for our daughter.

Yours sincerely,

Patrick and Veronica Miller

cc: Mr. Jackon Davies
        Principal, Couchiching District Secondary School

Commendation 4: Award Nomination - Corporate

(Award Commendation normally printed on corporate memo letterhead stationery)

CONFIDENTIAL MEMORANDUM

Date: September 12, 20xx

From: Peter Hartley

To: Ken Handler
            Executive Director

Subject: <u>Nomination – Barbara Meltzer– Roderick Shore Award, 20xx</u>

The purpose of this is to submit the name of Barbara Meltzer as a nominee for the 20xx "Roderick Shore Award for Excellence in Aircraft Accident Investigation."

Since July 20xx when Barbara first joined this agency, she has consistently demonstrated her superior skills, abilities, and professionalism as a member of the aviation accident investigation group. I believe that she is a shining example of everything that is signified by the "Roderick Shore Award", and she should be given the award for 20xx. I will briefly summarize my reasons for my nomination:

- Barbara was instrumental in leading a successful search and recovery effort when the Air Orion B-707 crashed into Lake Ontario in December 20xx.

- As investigator-in-charge of the Air Orion investigation, she has set a new standard for applying project management techniques to a major accident investigation.

- Barbara's performance in dealing with all parties involved has been exceptional. These parties included: next-of-kin, the media, police forces, interested parties, and other government agencies. She is clearly a gifted communicator and negotiator.

- She and her team of investigators managed to produce a comprehensive draft report for Board review within 15 months of the accident date. As you know, this is unprecedented for a major investigation and has set a new standard for this agency.

- Even though she was pre-occupied with the Air Orion investigation, Barbara managed to make significant contributions to the drafting of the Board's new "Investigation Policy".

I'm sure you will agree that Barbara's contribution has been outstanding and exemplifies the qualities of excellence, and professionalism that are embodied in the Roderick Shore Award.

I look forward to our discussion of nominees at next week's Management Council Meeting.

Peter Hartley
Director, Investigation Operations

Commendation 5: Community Service - Volunteer

(Commendation from a private citizen normally printed on standard stationery)

435 Fourth Ave.
Ann Arbor, MI
49637

October 15, 20xx

Mr. Albert Wolfson
Executive Director
Citizen's Volunteer Center
748 Center Street
Ann Arbor, MI, 49652

Dear Mr. Wolfson:

I am writing this to you on the advice of the Mayor's Office. My initial inclination was to write there but when I called, they told me to send my letter to you.

The purpose of this letter is to inform you of one of our citizens who I believe should receive some special recognition for dedicated community service. The person to whom I refer is Elizabeth Samuelson. I believe you know Elizabeth in your position as volunteer coordinator.

I have known Elizabeth for more than 20 years as both a friend and a neighbor. During that time I have seen her work tirelessly on scores of community projects and committees. As far as I know, she has never received any kind of recognition from the community for her work. I believe she should be given some sort of special recognition.

Just last year alone, she worked on at least six different community projects including the Park Renewal Project, and the Heritage Homes Fund-Raising Banquet. I believe she is also a member of a number of ongoing committees including the Library Restoration Committee and the Citizens for Literacy Committee. In addition, she is a weekly driver for the Meals-on-Wheels program and she visits patients at the palliative care unit at the hospital on a regular basis.

I'm sure that if you check with a few of your staff members who have been there over the years, they will confirm Elizabeth's community service record and probably be able to add many examples to the few I have given here. (By the way, Elizabeth has no idea that I am writing this. She is a low-key, humble person, and if she were aware of it, she would not allow me to send it.)

I have thought about writing such a letter about Elizabeth a number of times in the past but just never took action. Then last week, I noticed in the paper that you sponsor an annual dinner at which dozens of awards are given out to people with much less community service than Elizabeth. In fact, I checked with her and she has never even been invited to that annual event!

I urge you to correct that omission now and take action to see that Elizabeth Samuelson is properly recognized for her many years of silent but outstanding service to this community.

Yours truly,

Linda Sullivan

cc: Office of Mayor John Everest

PERFORMANCE EVALUATION LETTERS

In general, these are usually detailed assessments of an employee's work performance as part of an organization's regular employee review process. Typically, they are written by the employee's supervisor and are attached to the individual's performance appraisal and placed on their personnel file.

The format and structure for this type of letter is more often than not dictated by the "employee performance evaluation system" or process that is in-place where the subject of the letter is employed.

It should be noted that not all employers require a "performance evaluation letter." In fact, in recent years the vast majority of employers have developed their own internal performance appraisal systems.

These systems usually make use of fixed pre-formatted performance appraisal "checklist" forms that break a person's job performance down into six to eight performance areas, or factors. Narrative is kept to a minimum and there is no requirement for a "performance evaluation letter."

Nevertheless, the use of "performance evaluation letters" is still quite widespread in the academic community. Accordingly, this section of the guide is primarily focused on the evaluation letters related to the field of academia, particularly at the college and university levels.

DRAFTING TIPS – EVALUATION LETTERS

For general letter writing tips please refer back to the section earlier in this guide titled "Letter Writing Guidelines." (see page 23).

Following are some tips to keep in mind when drafting a performance evaluation letter about someone:

Stick To the Facts

Limit your comments to specifics, and avoid subjective statements without supplying substantiation. It is not good enough to simply say "Professor Sample's classroom was sub-standard". You need to substantiate such a statement with some facts or specific examples.

For example, "As evidenced by student evaluations, classroom audits, and review of teaching materials, in my judgment Professor Sample's classroom teaching falls short of the standard of professionalism expected in this department".

If you can't substantiate with some corroborating evidence, you're better off to not make the statement at all. If there is an area of weakness, but you can't really prove it, take measures to document it during the next session so that you can deal with it in the next performance evaluation.

Try To Keep It Short

There can be a tendency to go on and on when writing one of these letters, forever explaining and qualifying the points that are being made. Repetition is redundant in a performance evaluation letter, unless what is being repeated makes a completely different point. In general, make a particular point once, in a clear and concise manner, and then move on.

Watch Your Language

Word choice is particularly important when writing performance evaluation letters. People tend to take these letters personally! After the letter is written, you will have to discuss it with the person being evaluated. Later it will be seen by other people during the performance review process, and it will end up on the employee's personnel file.

Choose objective, neutral words and phrases. Avoid anything that is emotionally-charged in any way. If someone is a satisfactory performer, but you personally can't stand the person for some reason, keep this latter point out of the evaluation letter.

Keep It Confidential

Because of their highly personal nature, performance evaluation letters should always be kept confidential during all stages of the drafting process. Only those people directly involved in the process should ever see one.

SAMPLE TEMPLATES - EVALUATION LETTERS

The following pages contain six (6) fully-formatted real-life performance evaluation letter templates written for various real-life working situations.

Although the templates included here relate to the academic field, the approaches and phraseology can be applied in any situation where a formal performance evaluation letter is required.

Notes Re: Sample Letter Formats

- Although the sample templates in this guide are based on actual situations involving real people, identifying details have been altered to protect privacy.

- All of the sample templates in this guide have been reduced in size slightly to fit the book's page format which allows for page headers and footers.

- The samples use font size of 11 points rather than the 12 points I recommend as ideal. Also, the top/bottom and right/left margins have been adjusted to fit the margins of this book. You may want to readjust them for an actual letter.

Evaluation 1: Teaching - Satisfactory

(Performance Evaluation Letter is normally printed on corporate letterhead stationery)

CONFIDENTIAL

April 10, 20xx

Professor Ralph Hewson
P.O. Box 9010
Postal Station UNI
Michigan Ave. West
Chicago, IL, 60601

Dear Professor Hewson:

This performance evaluation letter covers your performance as a tenured associate professor from April 1, 20xx to March 31, 20xx. This evaluation focuses on three major areas weighted as follows: teaching (65%), research (25%) and service (10%). It also takes into consideration your responsibilities and obligations as a faculty member.

Teaching
Your teaching performance has been evaluated as average by both your peers and students. Based on my observations of your teaching methods and instructional materials during my periodic audits, I agree with that assessment. This average rating is partially offset by the great success you have achieved in developing a reputation as a preferred advisor. I congratulate you on your success in this area. In addition, you deserve the appreciation of the entire faculty for doing the lion's share of the work in preparing the instructional materials for the new first-year mandatory courses.

Research
I note that the majority of your research activities are directly related to new communication techniques and technology. I am also aware that two of your articles were published this past year by a very prestigious journal in the academic communications field. You have also presented a number of excellent papers on this subject. The CanComm Award that you received last year continues to give you many opportunities to showcase your work. It also reflects well on the department and the school as a whole.

In the coming year, I urge you to continue to submit papers to refereed publications and to continue to build up your dossier in your own field, rather than focusing solely on communications technology. Overall, I find your performance in research to be above average.

Services
In general, your governance and service activities continue to be satisfactory, and consistent with your departmental assignments. However, I have noticed a decline in this area during the past year, when compared with previous years. Your level of participation has fallen off and you avoid commencing new initiatives. I encourage you to get more involved. I suggest you accept the chair of the search committee as we try to fill the three vacant positions in the department. This would be an excellent concrete demonstration of your commitment to the service aspect of your position.

Overall
In summary, for reasons indicated in the foregoing, I have assessed your overall performance this past year to be satisfactory, but with improvements needed. I would like to see you focus more on diversity in both your instructional program and in your research endeavors. I encourage you to make a point to observe some of your colleagues in action and see how they conduct classes and keep the students engaged. As noted, I expect you to take a pro-active role in filling the vacant positions in the department.

In closing, I am pleased to recommend your appointment for the coming academic year. Your provisional schedule for next year is attached to this.

I would ask you to please sign and date this letter where indicated below after we have had our discussion on its contents.

Sincerely,

Agnes Ledbetter, Ph.D.
Department Head

By my signature, I acknowledge having discussed or reviewed the above letter of evaluation. I have __ have not __ attached additional comments with the intent that they become a part of my personnel file.

_____
Signature of Person Evaluated

cc: Dean M. Hughes
        Personnel File – R. Hewson

Evaluation 2: Teaching - Excellent

(Performance Evaluation Letter is normally printed on corporate letterhead stationery)

<u>**CONFIDENTIAL**</u>

April 30, 20xx

Professor Marilyn Chang
Performing Arts Center
Lakeview Campus
Berkeley, CA, 94705

Dear Professor Chang:

This letter constitutes the formal written evaluation of your assigned duties and professional responsibilities for the past academic year. This review is based on: your performance of assigned and associated duties and responsibilities, reviews of recent creative works you have completed, results of the Tenure Assessment Committee's annual review of the student teaching evaluation results, and on my own review and observations of your assigned and related duties and responsibilities. The three areas of evaluation are: teaching (80%), creative activity (15%), and services and administration (5%).

PERFORMANCE EVALUATION

Teaching and Instruction (80%)
As in previous years, your teaching continues to be excellent, both in the classroom and in the performance studio. Members of the Tenure Assessment Committee who have observed you in action have noted that you devote a great deal of personal effort and interest to your daily teaching activities, and to your students. Not surprisingly, student evaluations of your teaching are very high in all categories rated. I have noted that you continue to be open and receptive to suggestions as to how your teaching can grow and improve. I am impressed by the ways in which you have responded positively to some of the suggestions made during our previous annual reviews. You continue to make an outstanding contribution to the teaching and instructional program of this department.

I note that in addition to your assigned teaching responsibilities, you serve regularly on a number of committees, and frequently present master classes at other institutions. Your colleagues find you to be frank and direct in discussions but always ready to listen to new and/or different ideas.

Creative Activities (15%)
Your creative achievements remain excellent in every respect. In the past year, you have participated in numerous faculty performances on campus and in the community. Your recital performances are regarded as excellent, both within the state and across the country. You have also been invited to participate in an international concert tour in recognition of your excellence.

In addition to your performing, you have also achieved excellent success as a composer. Your compositions include six commissioned pieces, as well as three other published works. I am especially pleased that your work is frequently performed by orchestras in both this country and abroad. This is an amazing accomplishment for someone in an early teaching career and reflects very favorably on the department and the university.

Services and Administration (5%)
You continue to be involved in departmental management through your work on the University's Equal Opportunities Committee and your participation on the New Talent Search Committee. Additional service activities include your participation in a department-sponsored seminar, a performance festival, your service on the Board of Directors of the Berkeley Symphony Orchestra, and your continued performance as a member of the Valley Youth Orchestra. I believe that your level of involvement in these various activities is outstanding.

You have now completed your second year of tenure-earning status. I find that your overall performance in all three areas of teaching, creative achievements, and services/administration continue to be excellent, and completely consistent with your assignment.

PROGRESS TOWARD TENURE

The excellent quality of your work has been recognized and commended once again by the members of the Tenure Assessment Committee. You continue to achieve distinction, both on and off campus as noted in reviews and comments made by your colleagues and others. I note that you continually strive for improvement in all areas of endeavor. You are well-liked and respected by your colleagues. Your ability to respond positively to constructive suggestions and learn from them is exceptional. I encourage you to continue with your efforts in all of these areas.

As you enter your third year of tenure-earning status, I would like to suggest some ways in which you should continue to strive for continued growth:

- Focus on being more selective and discerning in the types and number of professional and creative activities that you engage in. Your shift in focus to more off-campus and regional activities last year had a positive impact and was wise. I suggest you prioritize a little more, in order to give yourself more balance and flexibility.

- Work on improving your awareness relative to how others view you in professional settings. Make sure that your interaction with colleagues will lead toward the accomplishment of your professional goals.

- Continue with your program to strengthen and expand your studio workshop series. The current enrolment of twelve students for next year looks promising.

- Continue on your path of composing and performing in ways that bring credit to you, the department, and the university.

- You should continue to refine your comprehensive teaching portfolio, which will be key in presenting a thorough profile of your teaching work when you apply for tenure and promotion.

- Continue to explore creative ways in which to develop new instructional materials for your classes, particularly for working with students of varying learning styles.

In summary, you continue to make wonderful contributions to our faculty and the department as a whole, and I regard myself as fortunate to be your colleague. Your hard work and exceptional commitment and dedication are appreciated by all whom know you. Accordingly, I am pleased to recommend that you be reappointed as Assistant Professor for the 20xx-08 academic year. Have a great summer and take some time for yourself!

The following is a list of your primary duties and responsibilities (scheduled and non-scheduled) for the 20xx-08 academic year. The list includes those other duties and responsibilities attendant to being a faculty member. Please note that due to the unpredictability of final student registration, your actual responsibilities during the coming year could be somewhat different from these. We often have to make adjustments in assignments after the teaching year has begun.

Teaching/Instructional Activities - 20xx-08 (provisional)
Performance Studio and Studio Workshop
Academic Classes in Music Theory
Composition Review and Critiques
Projects and Courses as needed
Thursday Afternoon Auditions
Faculty Meetings

Creative Activities and Services
In addition to teaching and instructional activities, it is expected that each member of the faculty shall engage in creative activities such as performance, composition, and research appropriate to the person's position. A form will be provided at the late-August faculty meeting, to be used for submitting a written statement of your plans for performances, research or other creative activities for the academic year.

Please review the draft of your annual performance evaluation letter and sign it the bottom where indicated if you accept the evaluation. You may attach any comments you might have for the record and return them to me no later than next Monday. If you disagree with the evaluation and/or would like to meet with me to discuss it or related matters, please make an appointment with me as soon as possible.

Sincerely,

Boris Kafelnekov, Ph.D.
Department Chair

_____
Faculty Member's Signature/Date

cc: Dean R. Gustafson
    Personnel File - M. Chang

Evaluation 3: Teaching - Borderline

(Performance Evaluation Letter is normally printed on corporate letterhead stationery)

CONFIDENTIAL

November 25, 20xx

Professor Ronald Bates
Arts Complex
UniCenter, Box 4250
Calgary, AB, T2E 5Y7

Dear Professor Bates:

This letter serves as the annual written evaluation of your assigned and related duties for this academic year at this University. It has been prepared in compliance with Article 43.7 of the directives of the University Board of Governors.

The evaluation of activity areas is weighted as follows: teaching (60%), research (30%), public service and administartion (10%).

Teaching and Instruction (60%)
Your teaching assignment for the past year (60%) involved courses at the undergraduate and graduate levels. You also served on three Masters committees (two as chair), and on three Doctoral Committees (one as chair).

As we discussed during our recent annual review meeting, I am quite concerned about a number of things that have transpired in some of your classes. These concerns have been reflected in low student ratings, and in complaints from students about your teaching practices. This is almost exactly the same situation that existed one year ago. There has been little or no improvement since then. These complaints have also been brought to my attention on several occasions this past year by graduate students. They include the following:

- You do not create a focused and orderly atmosphere in your classes. This is reflected by the absence of continuity and consistency in your class lectures. The haphazard way in which you follow up with students on their homework assignments has also been criticized. As I pointed out last year, more than once in the past few months, you need to focus more and organize your teaching approach.

- Your classes frequently do not begin on time and you do not provide appropriate instructional materials. These concerns were strongly reflected in student evaluations. I raised these points with you in your annual review last year, as well as in a series of conversations we had since the winter study break.

- You do not regularly attend academic meetings during which matters relating to the continuity of the academic program and related issues are discussed and decided upon. Numerous of your colleagues have stated that, when you do attend, you tend to be negative and are not particularly collegial in your interactions.

- You do not provide regular and consistent formal assessments and feedback to students about their progress and how they can work toward improvement.

- Your overall failure to perform your duties properly is clearly evidenced in the low student ratings of your teaching performance over the past year. Your ratings in all of your fall-term courses were extremely low. In the "Overall" category, yours were among the lowest of all faculty ratings during the same period.

With reference to the foregoing, and as we discussed when we met recently, I consider your teaching performance to be unsatisfactory. Accordingly, I must formally advise you now that if I do not see significant improvement in the areas discussed in this letter over the coming year I will have no choice but to but to take the actions I discussed with you.

I strongly urge you to contact the Department of Instructional Support for assistance in analyzing some of your lectures. In addition, I plan to visit your classes unannounced on at least three occasions during the coming winter and spring terms. My observations, coupled with your own analysis, should give you plenty of ways to improve your classroom teaching and demeanor. You might also want to ask one or two of your senior colleagues to observe you in the classroom and to provide you with constructive feedback which you can put into action.

Research And Scholarly Achievements (30%)
You have made several achievements in the area of research and scholarship (30%) over the past year. Most notable was the publication of your research in one article in a refereed journal, and your presentation of same, at a national conference. However, I find this level of productivity to be marginal considering your status as a Full Professor.

I expect you to increase your research and scholarly output significantly during the coming year. If your output in this area does not increase to my satisfaction, the percentage of time assigned for research next year will be decreased and an additional course will be added to your teaching load in future terms.

Public Service and Administration (10%)
Your public service and departmental administrative activities (10%) were borderline satisfactory over the past year. These included limited contributions and sporadic participation on the Departmental Library Acquisitions Committee. In addition, you participated in only 60% of the sessions of the International Scholarship Review Committee. I would also expect you to be more active in one or more of the professional organizations related to your specialty discipline.

It has come to my attention that, in addition to your specific assigned duties, you have also been involved in activities outside of the University that could be deemed in conflict. For example, your extensive consulting activities which were never reported and reviewed in accordance with the University's conflict of interest policy, appear to be in blatant conflict to the guidelines.

In light of the foregoing assessment of your academic and related activities as a faculty member, I find your overall performance over the past academic year to be unsatisfactory and in need of significant improvement.

Following are your assignments for the Winter and Spring Semesters:

- Three (3) Academic Courses – Please note that due to the unpredictability of final student enrollment, your actual assigned responsibilities during the coming year could be somewhat different from the above. It is often necessary to make adjustments in assignments after the year has begun.

- Departmental Committees as elected/assigned during the session.

- Advising students as required.

In addition to teaching and related activities, it is expected that each member of the faculty will engage in: research and scholarly activities appropriate for the individual's position, service to the Department and/or University, and performance of those other duties and responsibilities normally expected of a member of the academic teaching staff.

If you agree with this evaluation, please sign and date it where indicated and return it to my office by this Friday. If you disagree with this evaluation and/or would like to meet with me to discuss it, please make an appointment with me as soon as possible.

Sincerely,

Brian Jamieson, Ph.D., MSc.
Department Chair

By my signature, I acknowledge having discussed the above letter of evaluation and request ___ /do not request ___ that the attached comments be filed with this letter in my personnel file.

_____
Signature of Professor Evaluated/Date

cc: Dean Z. Ayoub
        Personnel file – R. Bates

Evaluation 4: Teaching - Unsatisfactory

(Performance Evaluation Letter is normally printed on corporate letterhead stationery)

CONFIDENTIAL

August 20, 20xx

Professor Judith Weiser
Campus Learning Center
1500 Heritage Circle
Cincinatti, OH, 45201

Dear Professor Weiser:

This letter is a draft of my evaluation of your annual and sustained performance as a tenured full-professor. The evaluations focus on your three major areas of assignment: teaching/instruction (60%) , research (30%) and service (10%), as well as on other related duties and responsibilities pertinent to your position at the University.

ANNUAL EVALUATION

Teaching and Instruction (60%)
As evidenced by my review of your teaching and instructional material and the results of your fall student evaluations, your teaching continues to be of superior quality. You have maintained a reputation as one of our most popular and effective teachers. Your involvement in teaching those difficult beginning level undergraduate courses is appreciated. You have also performed well in your other teaching-related activities and assignments, including your development of unique new teaching methods and aids.

Although your teaching evaluations remain above average, there has been a spate of recent complaints from a number of students about your conduct in the classroom, i.e. short-tempered behavior, use of inappropriate language, tardiness, being preoccupied, and not being prepared for class. We discussed these issues recently and you have indicated that you understand the concerns and will address them.

Research (30%)
Your activities for the past year in grant-funded research have been very disappointing. I realize that because a few of your graduate students are completing their work, it has been a busy time for you. Nevertheless, by failing to provide the required grant reports when required, you have missed the opportunity to have your current grant considered for continuation. Furthermore, you have not submitted any other grant proposals. Consequently there is no support for your graduate students. In addition, your publication activities have virtually ground to a halt. For these reasons, I consider you performance in your research to be unsatisfactory this year.

Service and Other (10%)
Your performance with respect to your normal departmental professional duties appears to also be at a standstill. Your attendance record at department committee and faculty meetings and seminars is spotty at best. It seems that in recent months you have almost isolated yourself from your colleagues and from the normal professional activities of the department. When you do occasionally attend, you appear to be confrontational rather than positive and productive. A number of your colleagues have expressed concern about this relatively sudden change in your demeanor. As I mentioned, in the previous section many of your students have also complained about your temper and sudden outbursts. We discussed this recently and you assured me that although you have recently experienced some personal problems, things are now under control.

Overall, with the exception of your classroom teaching, I believe your performance of the responsibilities to you as a faculty member are less than satisfactory this year, particularly the most recent term. This last period certainly does not reflect your normal performance. In summary, based on your annual report, student evaluations, peer reviews, and my own observations, it is my view that your overall performance is below average and in need of immediate and significant improvement.

SUSTAINED PERFORMANCE EVALUATION

With respect to your sustained performance over the past six (6) years, I find your overall performance to be satisfactory, excluding this past year. Your classroom work has always been your major strength, and I continue to view it as excellent. Prior to this past year, your research activities were always excellent. I realize that you have had some personal challenges and problems recently, which have degraded your performance. You have indicated that you are taking action on these and expect things to improve in the near future. As we discussed, in order to give you time to regroup, your classroom teaching assignment will be reduced this year. This will give you some extra time for course preparation, to conduct library research, and to revitalize your research program. We also discussed the possibility of you utilizing some of the on-campus resources such as life management counseling to help you deal with some of the problems you face. I trust you will avail of some of these programs during the coming year.

Please regard this as notification of your reappointment for the next academic year. Your tentative teaching assignments for next year will be two sections per term of your regular course in Literary Classics (50%). Your research assignment will be increased to 45% to give you more time to conduct research and develop and submit grant applications. In addition, your normal service component will be reduced (5%). As we discussed, I am hoping that you will also agree to chair two proposal review committees, as well as serve on the scholarship review committee. In addition to these assigned duties, you are responsible for other duties and responsibilities normally expected from a member of the faculty that are related to your employment at the university.

If you agree with this evaluation, please sign and date it where indicated and return it to my office by next Monday. If you disagree with it and/or would like to meet with me to discuss it, please make an appointment with me for early next week.

I wish you success in the coming year. Please let me know if I can assist you in any way.

Sincerely,

Marisa Thomasino, Ph.D.
Department Chair

By my signature I acknowledge having discussed the above letter of evaluation and request ___ do not request ___ that the attached comments be filed with this letter in my personnel file.

_____
Faculty member's signature/Date

cc: Dean L. Rassmussen
        Personnel File – J. Weiser

Evaluation 5: Teaching - Request for letter

(Performance Evaluation Letter Request is normally printed on corporate letterhead stationery)

November 24, 20xx

Dr. Simon Braithwaite
Professor, English Literature Studies
Department of English Literature
Arts and Sciences Building
7500 Edgewater Way
Portland, OR, 97213

Dear Professor Braithwaite:

I would like to thank you in advance for agreeing to draft a letter of evaluation for the tenure of Dr. Purie to the position of Full Professor in my department. To guide us in evaluating Dr. Purie's professional performance, we are seeking your considerd opinions as to the quality of the contributions that he has made in his field.

There are a number of factors that will be taken into consideration when deciding whether Dr. Purie should be promoted at this time. Although crucial to our evaluation, your input will be one of a number, so our decision will not be made on your input alone. We have contacted you as expert in Dr. Purie's field and ask you to please limit your comments to issues related specifically to your professional discipline. Here are the specific aspects that we would like you to address:

1. Provide background as to how long, and in what capacity you have known the candidate.

2. Your professional opinion as to the the quality, originality, and significance of the candidate's published works, with emphasis on the more recent work.

3. Your overall assessment of the candidate's national and international standing relative to other outstanding individuals in the same field at a similar stage of development.

Enclosed, you will find copies of various documents relative to your review. I realize that your review of these will be somewhat time-consuming. Nevertheless, I am sure that you are fully aware of the need for such rigorous assessments in order to maintain standards in the field.

During the evaluation process, your letter will be kept on a confidential file only available to members of the Evaluation Committee and limited academic support and administrative staff. However, as in most jurisdictions, the law requires that at the end of the review process the letter must be placed on the candidates's confidential personnel file, which will be available to him.

The Committee requires your letter of evaluation of Dr. Purie by January 15, 20xx.

On behalf of the university and the faculty of this department, I sincerely thank you for giving your professional time to this important matter. Don't hesitate to call me at (503) 287-5378 should you have any questions.

Sincerely,

Gerald Otterson, Ph.D.

Enclosures

Evaluation 6: General Surgery - Residency

(print Performance Evaluation Letter on institution letterhead paper)

March 15, 20xx

Mr. David Schwartz
Assistant Dean, Administration
East Coast School of Medicine
375 Oceanside Road, Rm. 450
Biddeville, ME 04010

Dear Mr. Schwartz:

Re: Performance Evaluation - Raymond Farmer

This Letter of Evaluation is provided at the request of Raymond Farmer, who is applying to your program for postgraduate training in General Surgery.

Premedical Education and Experience
Raymond grew up in Bangor, Maine. He did his undergraduate work at East Coast University where he received a BS degree in chemistry in May 20xx. Following graduation, Raymond spent a year working with a pharmaceutical company before entering medical school. He then entered EC University Medical School in August 20xx. He has completed the required curriculum with only the usual summer break between the first and second preclinical years, and is scheduled to graduate in May 2011.

Preclinical Record
Raymond passed all first-year courses without problems. His performance on the graded courses in the second-year was well above average. He earned the highest grade on 9 of the 16 graded courses and a high pass on the remaining 7 graded courses. He also achieved a score of 245 on the USMLE Step I examination in June 20xx.

Clinical Clerkships (Chronological)

Integrated Surgical Disciplines: Honors (H=48%; HP=48%; P=7%)
Tutorial: "Raymond was a good communicator; performing well on prepared presentations, and contributing to group discussions." General Surgery: "Raymond turned in an excellent performance on his first clinical rotation. He is bright and pleasant to work with. He worked hard and conducted himself in a courteous and professional manner. He evaluated patients in clinic in a timely manner and kept well-organized histories. He was very keen and helpful in the operating room." Ophthalmology: "Raymond was an excellent student. He is smart, diligent and reliable, and has a pleasant demeanor." Orthopaedic Surgery: "Raymond was an above average student. He will make a fine house officer."

Neurology: Honors (H=37%; HP=54%; P=9%)
"Raymond was a pleasure to work with in the neurology clinic. His presentations were well thought. His level of knowledge was excellent. He was an enthusiastic and helpful team player. He was well-liked by the patients. He will succeed in any clinical field he chooses." Raymond attained the 94th percentile on the NBME exam.

Psychiatry: High Pass (H=28%; HP=65%; P=4%)
"Raymond was hard working and helpful. His bedside manner was exceptional. His level of knowledge was also outstanding. He played an important role in the care of the patients assigned to him. He is a pleasure to work with and was well-liked by both staff and patients. His oral presentations were well organized and thorough." Raymond scored at the 95th percentile on the NBME exam.

Internal Medicine: High Pass (H=55%; HP=46%)
"Raymond is very hard working and contributes well to the team. His knowledge base is excellent and he uses each patient encounter as a guide for further reading and follow-up. He is a conscientious and dedicated student who showed dedication to every patient assigned to him. He was also very active in presenting articles and he read extensively. His level of knowledge is outstanding and he applies it at the bedside appropriately and well. Raymond will enjoy much success in the future."

Women & Children's Health/Pediatrics: Honors (H=37%; HP=60%; P=7%)
"Raymond's performance this month was exceptional. He was confident and had strong clinical skills. He has an excellent rapport with his patients and makes the extra effort to understand all aspects of their care. He is a bright and committed team member who cares about his patients. He developed excellent treatment plans for all of his patients and followed up on all assigned tasks. Overall, he is an outstanding medical student."

Fourth Year Electives

Cardiothoracic Critical Care: Honors
"Raymond is an enthusiastic medical student with a pleasant personality. He was always well prepared and his presentations were thorough. He worked hard and functioned as a full member of the team. He was totally involved in the care of his patients. He is a bright and willing learner and should be commended for a job well done."

Honors Medicine: Honors
"Raymond displayed outstanding knowledge and judgement throughout the rotation. His presentations were excellent and reliable. He functioned easily as an intern and will make an excellent resident. He has a great future. It was a pleasure working with him."

Research and Extracurricular Activities
Raymond spent the summer break between his first and second years of medical school conducting clinical research in the Department of Surgery. During his preclinical years, he was deeply involved with community outreach programs. He volunteered with the STOPS program (Students Teaching Obesity Prevention to Students), and coordinated that program during his second year. In addition, Raymond was a regular volunteer at the Downtown Medical Center, a free clinic run by medical students. He was also a member of the Surgery Study Group during his preclinical years.

Raymond enjoys hiking, mountaineering and music. He plays saxophone in a local student jazz band.

Summary
Raymond is a mature, well-rounded and bright young man. He has proven himself to be a hard working and capable medical student with the skills to be an excellent clinician. He is a very personable person who is dedicated to patient care and committed to a career in general surgery. Raymond's overall performance in the three years of required courses at East Coast University, Medicine places him in the top third of his class. He is therefore recommended to you as an excellent candidate for residency training.

Sincerely,

Marion K. Stremer, MD
Assistant Dean, Career Placement

BONUS CHAPTER!

COLLEGE ADMISSION ESSAYS

This guide wouldn't be complete in covering recommendation letters related to college and university admission without also addressing one of the most critical documents that must also accompany the application for admission – the college admission essay.

As entry into college and university programs becomes increasingly competitive, the importance of a well-written admission essay or essay set cannot be over-emphasized. These essays are normally required for consideration for admission into both the graduate and undergraduate programs, although the specific requirements may vary at each level.

Aside from the standard academic performance metrics, these admission essays are the one chance that a student has to show who they are, what they've done, and how they can express themselves. They go beyond the actual academic achievement to show a little bit about the actual "person" behind the application.

It's a chance for the student to show their human side to the officials that review admission applications.

ADMISSION ESSAY TERMINOLOGY AND REQUIREMENTS

The terminology used to describe these documents can be very confusing at times. My research has revealed that there are a number of different terms that are widely used to signify the generic "college admission essay." These terms vary from country to country, and often from school to school, and sometimes within the same country.

Here is a list of the most common terms that are used to describe the "college admission essay":

- Admission essay

- College essay

- Graduate admission essay

- Personal statement

- Statement of purpose

Specific requirements for what must go into the admission essay will vary for college to college. Sometimes, the requirement will be to write a one to two page essay on the topic of your choice. Other times, you may be asked to choose an essay topic from a list of pre-specified topics.

A common approach is to ask an applicant to write an essay that answers a specific question. For example, the question could be "Define your five (5) most dominant personality characteristics and describe how they will affect your future academic and professional careers?".

In most cases, unless otherwise stated, your resume should range in length somewhere between 600 and 800 words (about a page and a half, single spaced).

A few universities may even require you to write a set of short essays on a list of pre-selected topics (e.g. Harvard).

Regardless of the specific requirements at a particular university or college, the overall approach to writing an admission essay is essentially the same.

ADMISSION ESSAY REVIEW PROCESS

Typically, the admission review process involves three levels of review: admission assistants, associate directors of admission, and the director of admissions.

Your application and essay will be reviewed first by the admission assistants. These people are normally recent graduates, perhaps four or five years older than you. If they like your essay, they will recommend it for further review at the next higher level.

The next level is the associate admissions officer. These are normally career employees at the college or university who have spent many years reviewing admission submissions and essays. If the associate admissions person likes your essay, they will pass it up the ladder to the Director of Admissions who will make the final decision.

Together, these three levels are often referred to as the "Admissions Committee."

Most colleges and universities receive thousands of applications beyond the number of students that they can admit in any given year. This relatively small admissions review group must review each and every application and read each and every essay for applicants that meet the minimum admission requirements.

Needless to say, during peak periods these people are overloaded with work and often have to read dozens of essays in one day.

This is why it is so important that you write an admission essay that stands out from the crowd. Your essay must grab the reviewer's attention in the opening paragraph. It must draw them in and interest them at once, and convince them to move your file on the pile designated "for further review" that will go to the next level.

If your essay makes it to the Director of Admissions level, there is a very good chance that your application for admission will be accepted.

WHAT THEY'RE LOOKING FOR

It's not just about the marks.

The people on the Admissions Committee are not looking for more than evidence of your academic achievements. Enough of that type of information is already included in the other sections of your application.

Colleges and universities are not just looking for academic geniuses these days. As society and the world change, these institutions are also seeking people with diverse backgrounds, experiences, and interests. They want people who will make unique contributions to university life, and eventually to society in general.

Your admission essay allows you to show another dimension of yourself that may not come across clearly in the rest of your application.

It's about you, the human being. The essay can highlight your personal strengths and unique personal characteristics and show you as a balanced and well-rounded individual, rather than just an academic achiever.

Don't underestimate the importance of the essay. All other things being equal, it is the essay that will determine whether you get accepted, or not.

DRAFTING TIPS – COLLEGE ADMISSION ESSAYS

As mentioned above, the specific requirements for essays will vary from school to school. However, the general approach and the basics for writing a good admission essay are essentially the same.

The following are 16 tips that will help you draft a better college admission essay:

1. Make It A Personal Statement

The admissions committee isn't looking for bureaucratic bafflegab or pie-in-the-sky platitudes. They want to know a little bit more about you, the person, the human being. Who exactly are you? What makes you unique? How do you see the world? What do you have to offer as a person? What will you bring to the academic and social life at the university? The general essay requirements and the questions are designed to elicit this kind of information from you. Make sure you don't blow this opportunity.

Here's an important word of caution. Don't forget that the overall purpose of the essay is for the readers to get to know you in the broader sense. Make sure your essay doesn't focus exclusively on one aspect of your personality or one life-changing event. Make sure you paint the bigger picture of you.

2. Write It For The Committee

When you are planning and writing your essay keep in mind at all times who will be reading it – the Admissions Committee. Try to picture these people in your mind's eye as you write. What would they expect? What kind of information will they be looking for? Are you presenting the material appropriately for this audience?

To get a better "feel" for the Admissions Committee it can be helpful to go to the school's Web site and spend some time in the Admissions area, seeing how they are organized, and viewing the names and qualifications of the staff listed there. This can make it easier to visualize the committee as you write your essay.

3. Keep The Language and Structure Simple

The most effective approach is to write the essay as you would speak. Imagine yourself sitting across a boardroom table from three of four members of the Admissions Committee. Write your essay just as you would speak in that situation – simple, direct, clear. Keep the paragraphs and sentences short. Breaking your essay into three or four sentence paragraphs will make it more pleasing to the eye and easier to scan.

As an editor, I can tell you that there is nothing worse than having to work on a dense page of run-on sentences with few paragraph breaks. You just don't want to go on reading. If you have good content, don't turn your readers off with poor sentence or paragraph structure.

4. Don't Repeat What They Already Know

Make sure that your essay doesn't just reiterate information that is already included in other parts of your application package. For example, don't spend time going over your academic achievements when they are covered in another specific part of the application.

For your essay, choose a topic or an angle that presents different information about you that can't be found elsewhere in the application.

5. Answer The Question

Many college admission applications will ask you to answer a specific question. Questions such as: "Describe a significant childhood experience and how that shaped you into the person you are today." Or, "Describe the one or two individuals who have influenced you thus far in your life and explain how that influence has changed and/or guided you."

If you get one of these questions, be sure that you answer it directly and specifically. When you are at the final draft stage it's a good idea to go back to the original question and make sure that you have answered it directly and completely.

Sometimes the question will be much more open-ended, such as, "Describe your personal characteristics and explain how they prepare you for admission into this program." Even with a broader question like this, it is easy to get off-course.

Don't write your essay about a tragedy in the family unless you tie it directly into how it shaped you into a prime candidate for the program to which you are applying.

6. Don't Try To Be Too Cute

Don't take your essay too lightly. Remember, these admissions committee people are faced with making difficult decisions that impact people's lives. A sense of humor can be great and there is nothing wrong with showing this side of your personality in your essay. Just don't go overboard. The readers of your essay are being asked to make a serious decision that will change your life. They need some serious straightforward material to do that.

Also, be careful using humor. A sense of humor is a very individualized attribute. What may be hilarious to one person can fall flat with another.

7. Target The School If Possible

Your ability to do this will vary depending on the specific form of your essay and/or the wording of your question. However, if you possibly can, weave the target school into your essay. First, go to the school's Web site. Take a look at the opening pages that give the general information about the school you are applying to. Look for statements like: mission, vision, guiding principles, educational priorities, etc. Letters from senior faculty members are often good sources of material for this.

If you read this material carefully you will spot certain words, phrases and/or ideas popping up again and again. These are the signposts of the institutional culture of that college or university. If you can possibly slant your essay to incorporate one or more of these themes, it will definitely add value to your submission.

For example, say that one of the overriding themes of the college you are researching is, for example, "… social and cultural diversity for the new millennium." See if you can find something about you or your experiences that ties into this theme. Perhaps you are from another culture. Or, maybe you have traveled widely or lived abroad and therefore have a better than average understanding of other cultures.

This is just an example. If you look carefully and think about it, there are probably a number of these or similar themes that you could relate to in your essay.

8. Focus On Your Uniqueness

You want your essay to somehow make you stand out among hundreds of others. To do this, your essay must be different. The essay itself can use a different approach (which can sometimes be risky) and/or it can highlight something particularly unique about you.

Think about what unusual experiences, interests, and/or hobbies you might have. Try to weave one or more of these into your essay in an interesting way. Think hard. Everyone has some unique and/or unusual characteristics and/or interests that make then stand out from their peers in some way.

You don't have to be a child prodigy who became a concert pianist to be unique. Maybe you have an impressive butterfly collection, or you are addicted to golf. Working these into your essay in an interesting and positive way will help your essay.

9. Tell A Story Of Personal Change and Growth

Essays that tell a story of how you have changed in a positive way due to one or more events in your life can be very effective. Or, it might have been another person who influenced you to change. Simply, describe that event and/or person and explain how it changed you for the better. A story of how you faced a personal problem or challenge and managed to overcome it, and how that changed you to become a better person can make for a powerful essay.

But, be careful to make sure that your essay is focused on you, and how you changed, and not on the life-changing event or person.

10. Write About You and What You Know

Sometimes there is a tendency to write beyond one's personal experience and knowledge in order to impress, or meet perceived expectations. This is never effective, and can almost always be recognized by experienced readers. Stick to what you know, what you've experienced, and where you have been. A simple story can still make a good essay if approached properly and written well.

Most importantly, be honest and be yourself. Don't try to invent a new you. Living a lie is just not worth it. If you did manage by some small miracle to get through the admissions committee without being identified as a fraud, it would just be a matter of time. Eventually you will get caught and be exposed, and then everything that you have accomplished up until that time will be in jeopardy.

11. Avoid Controversial Subjects

Your college admission essay is not the time to get up on the soapbox and try to convert people to your point of view. Members of the admission committee are from all kinds of backgrounds and socio-economic groups.

So, regardless of your personal beliefs in certain sensitive areas, do not write on a subject that might offend one or more of these people. If you do, you will be shooting yourself in the foot. Doing this will definitely not help you get accepted into the college or university of your choice.

Specifically, avoid writing about politics, religion, or controversial issues such as the legalization of marijuana, abortion, and euthanasia.

12. Provide Supporting Details

To make your essay interesting and to give it more credibility, make sure that you provide supporting details. Involve the reader using colors, smells, visual descriptions, and feelings. Show that you've "been there and done that." This will engage the reader and will make them relate to you more as a real person, as opposed to just reading neutral, non-involving words on a page.

For example, the sentence "I can still smell the sandalwood smoke that greeted me when I stepped off the airplane at 2:00 a.m. in Bombay" will have more impact than "We arrived in Bombay by plane at 2:00 a.m."
Remember, you're telling a story.

13. Draft and Then Re-Draft

A good essay should go through at least three or four drafts. This is all part of a natural process that will eventually arrive at a final product. The more time you give yourself to go through this drafting process, the better. If you can do it over a two or three week period, with 3 or 4 day gaps between each draft, that's great. Each time you look at it you will see it from a different perspective until you have fine-tuned it into what you consider a good final draft.

14. Proof Read and Edit Carefully

Remember that you will be submitting your essay to an academic institution and it will be reviewed by well-educated people with above average writing skills. Accordingly, there is no quicker way to lose your credibility than to submit an admission essay with typos and/or grammatical errors.

Use the spell-checker and grammar checker included with your word processing software. If you don't have one, there are plenty available for free online. Just go to your search engine of choice and enter "free spell checker" or "free grammar checker" to find one.

Remember, although spell-checkers are great, they don't catch everything. For example, I often reverse the letters in certain words when typing quickly. i.e. "form" instead of "from." As far as a spell-checker is concerned, these are both valid words. Some grammar checkers will flag this as being out of context, but you can't always count on that.

The only way to be sure in the end that everything is fine, is to have someone with good spelling and grammar skills do a final check.

15. Get Feedback From Others

Late in the drafting process, when you are sure of the approach and content and are just in the fine-tuning stages, it is a good idea to ask someone you know and trust to read the essay over and give you their feedback and/or impressions. The two main points you will want to derive from their feedback are whether your essay is an accurate reflection of you, and will the writing stand up? Ideally, this person will be someone who knows you reasonably well and has good writing skills.

In fact, it is a good idea to get a couple of opinions from people with different backgrounds and qualifications who will see the essay from very different perspectives.

16. Sleep On It One More Time

After you have arrived at the "final draft" of your essay, put it aside and don't think about it, at least for overnight. Even if you aren't thinking about it consciously, your sub-conscious mind will still be working on it. I know this very well from personal experience. Whenever I write something for a client, I try to leave enough time so that I can "sleep on it" and then do one final pass the next day, before I submit it.

No matter how hard I have worked at drafting the document, I invariably find a few final adjustments to make the next day. Without exception, these "adjustments" always improve the final product.

SAMPLE TEMPLATES – COLLEGE ADMISSION ESSAYS

As I mentioned at the beginning of this chapter, the terminology used to describe these admission documents can be somewhat confusing at times.

There are at least five different terms that are commonly used to describe essentially the same thing; the college admission essay. These terms vary from country to country, and sometimes from school to school within the same country.

As already explained, these generic terms for the college admission essay are: admission essay, college essay, graduate admission essay, personal statement, and statement of purpose.

Again, the specific requirements for what must go into the admission essay will vary from college to college. Sometimes, the requirement will be to write a one to two page essay on the topic of your choice. Other times you may be asked to choose an essay topic from a list of pre-specified topics.

Nevertheless, regardless of the specific requirements at a particular university or college, the overall approach to writing an admission essay is essentially the same.

The following pages contain some typical college admission essays. These essays are based on real people in real-life situations but names and specific identifying details have been changed to protect privacy.

Each of the five (5) essays that follows is a response to some of the most common questions asked in college admission applications.

After carefully reviewing these, even if your question(s) is different, you should have a good idea as to how to draft your own essay.

Admission Essay 1: Life-Changing Experiences

Describe one or two significant childhood events that have had a lasting impression on shaping the person you are today.

I'm not sure whether it was that day at the swimming pool when I was 10, or the week that it poured rain when I was 12 that had the biggest impact on the person I have become today. What I do know is that both of those events were important moments of realization for me.

There I was, red-haired, freckle-faced and skinny-white, standing beside the swimming pool with the other boys in my class. All twenty of us were standing there, shoulder-to-shoulder, each one waiting his turn to dive in head-first, swim underwater to the other side of the pool, resurface, and then climb up on the opposite side. That was the simple object of the opening-class-exercise that our burly and gruff male instructor had assigned to us. There was only one problem with this picture. I didn't know how to swim!

But they started anyway. One-by-one; diving in, swimming across the pool, and then jumping up and out. I was second last in line. As I watched the others take their turns, I was filled with fear, wanting to just run up to the instructor and say to him "I can't swim". But for some reason, I just couldn't do that. When my turn finally came, I dove head-first into the water as if I knew exactly what I was doing. Then I proceeded to start drowning. Fortunately, there was a life-saving pole nearby and my instructor managed to fish me out as I choked and spluttered mouthfuls of chlorinated water. Humiliation isn't a powerful enough word to explain what I felt at that moment. Profound shame might be a better description.

This was a clear demonstration as to how far I would already go by the age of 10 years, just to please people. What that incident really boiled down to was the fact that I would choose to look death in the face rather than admit that I couldn't do something. I later realized that I would go to such lengths particularly to please older men, especially if the older man happened to resemble my father in any way. Not surprisingly, the swimming instructor was a younger version of my dad. Why exactly that dynamic existed, is the subject for another essay

But you know what? I did learn to swim that summer. After that incident I somehow managed to screw up my courage, go back to the swimming class, and keep on diving into that water. In that swimming pool, I learned that by facing my fears head-on, I could overcome them.
A couple of years later I was the star pitcher in our suburban baseball league in Dartmouth, Nova Scotia. Near the end of the season the coach asked me if I would like to be the starting pitcher in a

special game that had been arranged against the all-star team from the "big city" across the harbor from us. This was a wonderful thing. It was a huge honor, and I had never been so excited about anything before.

The game was to take place the following Saturday. I could hardly wait and had never anticipated something with so much excitement. Saturday came and it was pouring rain. The coach called and said that the game was post-poned to the next day. Sunday came and it was still pouring. The game was re-scheduled for the following Wednesday evening. On Wednesday, it poured all day once again, and the game was re-set for the next Saturday. I believe that we had rain for 30 straight days that summer. It was a record. After Saturday's rain-out I never did hear from that coach again. Somebody said he had gone away on vacation. I felt an empty hole in the pit of my stomach. It stayed right there with me for the rest of that summer, and beyond.

That experience taught me that you can never count on anything for sure. You can plan, and you can prepare, but for the actual event you just have to wait until you get there and see what happens when that day comes.

Now as I seriously contemplate going to college I realize the importance of these two lessons that I learned almost a decade ago. Lesson number one: I can face my fears, and if I do, I will overcome them. Lesson number two: I can do all of the preparation and planning necessary for an event. But I won't worry or obsess about what is going to happen in the future. When the big day comes I'll accept whatever it presents me with.

Admission Essay 2: Travel and Cultural Diversity

Explain how your international travel experiences have affected your view of the world and how they have impacted you as a person?

From as far back as I can remember I have been shuttled back and forth between countries. Later on, it was also between houses, and then finally, between cities. Little did I know at the time, the impact that this living out of a suitcase would have on me as a person, and how it would change my understanding of people, and the world.

My dad and mom met in Canada, were married in India, and then settled back in Canada. After a few years in Canada's capital city, Ottawa, my dad joined a special project team in St. John's, Newfoundland. That's where I was born. So, that makes me a Newfoundlander with an anglo-Canadian father and an indo-Canadian mother. They don't have a name for that combination yet.

After returning to Ottawa, my parents separated and eventually divorced when I was four years old. From that point on, I lived full-time with my mother and stayed with my dad every second weekend, as well as during vacation periods. For a few years, I also stayed with my dad one night each weekday. We did a lot of shuttling back and forth.

Ever since I was a baby I was taken to India by my mother almost every Christmas season to visit my grand parents and cousins. That trip is something I really looked forward to every year. India is such an exciting and exotic place, and I loved visiting my relatives there. Most years, I would leave for India on Boxing Day so that I could also spend time with my dad and his family in Canada on Christmas Eve and Christmas Day.

Once I turned 11 years-old, I was able to fly to India alone and spent most of my summer vacation each year with my grand parents and cousins in New Delhi. Occasionally, I would miss a few days of school in early January because of the India trips. To compensate for this, I would write essays about my experiences in India and in Europe during my travels. I would hand these in as assignments when I got back to school in Ottawa, Sometimes I was even asked to read them to the class.

So, in many ways I have experienced the best of two worlds, India in the East, and Canada in the West. I am fortunate enough to have extended families in both places.

Then, when I was 13 years old my dad moved from small city Ottawa to the big city of Montreal. Since then, I have bused to Montreal every second weekend to see my dad. I have always loved my time in Montreal. It has the feeling of a large vibrant international center that is dripping with multiculturalism and ethnic diversity. I also get to practice my French whenever I'm there.

None of my friends in high school have traveled quite as extensively as I have. Sometimes in class, or even just in casual conversation, I realize how much I have learned in my travels and how they have allowed me to see the "bigger picture" as my dad calls it. I find that my friends sometimes have trouble understanding the differences of other cultures. I know that this is because their only reference point is their own culture. On the other hand, although I don't pretend to know a lot about cultures other than in Canada and India, I think that knowing that there is another dimension leaves me more open to whatever differences another culture may have. I have learned that these differences do exist and must be accepted as they are, with no judgment by one group or the other.

I have no doubt that my multi-cultural background, coupled with my traveling experiences, have created my current interest in pursuing studies in world affairs and international development.

The way I see it, we're all together here, co-existing in our diversity, on this beautiful planet of ours. So, let's just try to be tolerant and accepting and enjoy each other's company. With the proliferation of the Internet and the global economy, the world is quickly becoming a much smaller place than it was during my parent's youth. Before long we're all going to have to learn to get along as one big extended family.

Admission Essay 3: Targeted University

Explain why you have chosen to apply for admission into this particular college (or university)?

I know it sounds a little strange, but I have wanted to go to McGill University ever since I was 10 years old.

I believe I picked up this notion during one of my semi-annual visits to my cousins' place in London, England. Often, during dinner conversation, the adult discussion would turn to where the kids (us) would go to college when they (we) got older. I don't remember much about those conversations except for one thing. Invariably, I would hear the phrase, "… in North America, it has got to be either Harvard or McGill!"

This was always stated quite emphatically, as if it was a fact, that there were no other choices. At the time, this didn't mean that much to me, but I stored the information away for future reference. Then one day in Miss Hickson's Grade 5 class we were asked to name the university that we would like to go to. I immediately blurted out "Harvard or McGill" believing I had pretty well covered all of the options. Boy, was I surprised when some of the other kids yelled out a series of strange names like U. of T., UCLA, Cambridge, McMaster, UBC, Queen's, Cornell, Western, and others.

Miss Hickson went on to explain that there were many universities and colleges (in fact hundreds) across North America. This was definitely news to me at the time. She then stated that Montreal's McGill University was considered by most people to be among the very best in Canada. Of course that confirmed everything that I had heard over the dinner table in London. So from then on, for me, McGill was the place to be if you were going to go to university in Canada.

After that, I became very McGill-conscious. My ears would perk up whenever I heard the name mentioned on the tv or radio. I would notice McGill references and articles in newspapers and magazines. McGill jackets would stand out in a crowd for me. At one point I even bought McGill track pants to wear to high school.

When I was 13 years old my dad moved to Montreal. This was a big deal for me at the time, because not only did it mean Montreal, an exciting new city to visit, but it also meant I would be close to McGill University. I would travel there by bus every second weekend to stay with my dad. On those weekends, my heart would always beat a little faster when we would go by the old greystone buildings of the McGill campus, nestled half way up beautiful Mount Royal in the heart of the city.

When I was 15 I traveled to Montreal with the Ottawa Lion's Club track team. The meet was held at McGill's indoor track and field center. I felt at home immediately. In fact, for me it somehow felt like coming home. The corridor walls were plastered with photos of McGill sports heroes of the past. For some reason I don't fully understand, I felt proud of these bygone stars, and somehow strangely connected to them.

As I got older, I came to appreciate McGill's reputation as an international center for higher learning. Many famous and accomplished people have taught there and/or graduated from there. World famous pioneers like Linus Pauling and Norman Bethune. There were personal and family connections to McGill as well. Stephen Leacock had taught there for many years while spending his summers in my dad's birthplace, Orillia, Ontario. My dad's favorite poet, Leonard Cohen had graduated from McGill.

During my last two years at high school I worked harder than ever, intent on keeping my average in the mid-80s to make sure that I would have a chance to get in to McGill when the time came. I really focused on my English literature and world issues courses since I was interested in applying for political science or sociology, eventually doing an honors degree in International Development.

As I came to know Montreal, I realized how McGill and Montreal fit so perfectly together - a world-class international city, hosting a world-class international university. As one walks up McGill-College Avenue approaching the McGill campus, one sees the university and its mountain reflected in the windows of the high-rise office towers, opposite. Each one, a perfect reflection of the other.

Now, as I contemplate the possibility of attending McGill as a student, a lump forms in my throat as I realize how close I am to achieving that childhood dream. If I am accepted, it will be my turn to contribute to the continued building of that institution's international reputation for excellence. It's an awesome challenge that I definitely want to be a part of.

Admission Essay 4: Career and Personal Goals

What are your career goals and why have you chosen that particular path. Where do you see yourself in ten years?

One of my best friends from when we were toddlers knew from the time she was eight or nine years-old that she wanted to be a doctor - no ifs, ands, or buts. Susan was going to be a doctor and that was it. The funny thing is, just last week Susan was accepted into her second medical school. I'm sure that she will do well in her studies and will become a great doctor.

It hasn't been quite so straightforward for me. In fact, to be perfectly honest I don't yet have any fixed career goals. But I do know some general directions that I want to go in, both professionally and personally. I just don't yet have a fix on a specific job or career path that I want to pursue.

My experiences both in and around high school have shown me some things I am good at and some things that I like to do more than others. For example, I love working face-to-face and shoulder-to-shoulder with other people, especially if we are working towards a common goal. I realized this when I worked as a co-organizer on a campaign to raise funds for cancer research. I adored the feeling of working with others for a great cause. I liked the concept of being able to help other people suffering or in need.

My work as a volunteer with handicapped and learning disabled children was also very rewarding for me. I enjoyed being around these kids and experiencing their innocence and their unconditional love whenever I would do the smallest thing to help them. These kids are not always easy to work with but the rewards can be great. It was nice to be able to make a small difference in someone's life every day, even if it might seem to be relatively minor stuff to some, in the big scheme of things.

I love writing, especially about current issues. Working as a feature writer on the school paper for the last two years of high school was like a dream come true for me. Since I had the "world issues beat" I was given carte blanche to focus on whatever I considered the more important issues on this planet right now. Pressing issues, like infant mortality, child poverty, and the general exploitation of children, especially in developing countries. I really enjoyed researching these subjects and writing articles about them. Again, by exposing these issues to a larger audience I felt I was helping people in suffering or need, at least indirectly.

This would explain why I have applied for admission into a general arts program with political science or sociology as my majors. I believe that during my studies in these areas I will develop a much clearer idea about which specific career path I might want to follow. In the meantime, I will continue to learn more about who I am and what this world is all about.

As for where I see myself ten years from now? Who can really know? There's no point in pretending that I know where I'll be or what I'll be doing. What I do wish though, is that ten years from now I will be happy and successful in whatever profession I will have evolved into during my university career. More than anything, I hope that at that time I will still be working, learning, and creating - and changing too!

I don't ever want to get "stuck" in one place in my development, either as a professional or a person. Ten years from now, if I can honestly say that I don't know what I'll be doing ten years from then, I'll know I'm on the right path. After all, I believe we're all here to keep on learning - forever.

Admission Essay 5: Social Issues and Concerns

Discuss some of the main issues that concern you at the local, national, or international level.

"You're seeing the glass as half-empty again" my dad would invariably say to me as soon as I would launch into one of my frequent tirades about some new social injustice I had just discovered in my high school World Issues class. "Don't forget to look at the other side of the equation" he would admonish me.

Even now as I look back at one of those discussions that occurred just six months ago, I realize how my thinking has changed. I think I'm starting to see the big picture more than I used to. I'm starting to realize that many issues are very easy to react to at the basic gut level, but when one looks at them more carefully the picture is frequently much more complex and the "right" answer is not always so cut-and-dried anymore.

Mind you, this doesn't apply to everything. Some things are just plain wrong. War for example. Both sides may have a legitimate case but murdering each other's citizens can never be justified. Sexual abuse of children is another. This is sick and inexcusable under any circumstances. Essentially, I believe that all forms of violence against the human race are fundamentally wrong.

While working on a couple of projects in my final year of high school I had a chance to do some literature research into a number of these issues. What I found was very interesting. I came to realize that there are many social issues and conditions that aren't quite as straightforward as the ones I mentioned above.

For example, most people are not aware that by far the largest killer of human beings on this planet is indoor air pollution, estimated to kill 2.8 million people every year, mostly in the developing world. That's right, I said million. Yet, when we read our newspapers and listen to our politicians we never hear about this problem. Instead they go on-and-on about things like mad cow disease and the West Nile virus. Yes, it is unfortunate that some people are dying of these. However, the fact is that these two conditions together have killed fewer than one thousand people worldwide, ever. Compare this with 2.8 million human beings killed annually by indoor air pollution that is directly related to ignorance and living in poverty.

Global warming is another issue that I think is very important but one that is also being distorted and blown way out of proportion by the media and special interest groups. The fact is, that even the scientists can't agree on exactly what is going to happen in the future, and their estimates are wildly different. So, I can understand the reluctance of the U.S. government and some others to resist immediate implementation of the Kyoto protocol. Instead, they believe in a phased adoption of the Kyoto measures because immediate implementation would cost billions of dollars and throw the economies of the major developed countries into chaos. This would benefit no one, and would cause more pain and suffering than the short-term effects of global warming.

There are a number of other global issues that also interest me, including: human poverty and starvation, environmental issues in general, and biodiversity.

An aspect of all of this that really interests me goes back to my dad's analogy about the glass. Is it half-full or half-empty? Why do the media, politicians, and special interest groups tend to jump on the bandwagon of negativity on just about every major social issue? Is this healthy? Maybe if we're always preparing for the worst case scenario all the time, we can't go wrong. But then, what about our tendency to focus on the wrong things due to all of the hype surrounding the "designer" issues of the day?

These kinds of questions fascinate me. I guess that explains why I was drawn to apply for admission into a political science program majoring in world issues. After that, I think would like to enroll in a masters program in journalism in which I could focus on covering issues related to the developing world and the future of the human race. I would like to be able to add some balance to that coverage.

COMMERCIAL WRITING SERVICES – COLLEGE ADMISSION ESSAYS

There are numerous websites offering various levels and types of service to assist people in the writing of essay admissions of every description. Some of these sites also offer other writing services for related documents such as recommendation letters and college application letters.

To view some of these sites just go to the search engine of your choice and enter the search phrase "college admission essay" or "personal statement essay" and you'll be overwhelmed by the number of results you get. Even if you take the time to investigate the top search results returned and ignore the weaker matches, you'll still end up with at least 10 to 15 services on your "short list".

So, I did some detailed research into what some of these services charge for their essay writing services. Hiring one of these services to edit and/or write your essay(s) can be a fairly costly proposition.

Here's a page I posted on one of my websites that summarizes what some of the major services charge.

http://www.instantcollegeadmissionessay.com/CostTable.html

As you can see from those numbers, the cost of writing and revising a few essays or personal statements will add up very quickly using these outfits.

So, if you're looking for some professional, yet inexpensive help with your college admission essay of personal statement, you might want to check out my ebook on this subject:

http://instantcollegeadmissionessay.com

ONLINE RESOURCE LINKS

I believe that if you faithfully follow ALL of the advice and information included in this ebook you will be able to successfully write any recommendation letter.

Nevertheless, even given all of the information provided here, there will be some people who won't feel entirely confident until they check out every possibility. So, those folks will continue to look for additional help and advice.

To save those people a lot of time and trouble I have spent many hours researching what additional online resources are out there that one may want to consult.

When I did a www.google.com search on the term "recommendation letter", over 725,000 results were returned. For "letter of recommendation" it was 2,030,000. For "reference letter" total results found were 1,400,000, and it was 269,000 for "letter of reference".

So, to get that list down to a meaningful number of sites containing quality information, I did a detailed review of the first 30 to 50 of the most relevant results from each search and then chose the one that I considered to be the most relevant and useful out of those.

I then listed those sites ranked in order of popularity according to their alexa.com traffic ratings on the day that I conducted this research. Alexa is a respected website ranking service that computes traffic rankings by analyzing the Web usage of millions of www.alexa.com toolbar users.

For a subject like "recommendation letter", the Alexa ranking, which is based largely on volume of traffic, can be considered a reasonable indicator of the relative quality of content on a website.

As you can imagine, researching and compiling this short-list of links took many tedious hours, so please take advantage of the fruits of my labors.

TOP 15 RECOMMENDATION LETTER SITES

The following links are what I consider to be **the top 15 recommendation letter writing resource and information sites** to help you find any additional resources if you feel you need additional help with your letter writing. When assessing the various sites I used organic search results only and avoided commercial websites.

As mentioned above, the sites are **ranked in order of popularity according to their alexa.com traffic ratings** on the day that I conducted this research.

Please note that at the time this research was conducted and this list was compiled, all links were tested and were found to be in good working order.

Recommendation Letter Defined By Wikipedia: The term "recommendation letter" is often used interchangeably with the term "reference letter"; however ...
http://en.wikipedia.org/wiki/Recommendation_letter

Good Letters of Recommendation: Whether you are writing or requesting a recommendation letter, you should know what to include ...
http://businessmajors.about.com/od/recommendationletter1/a/GoodRecLetters.htm

How To Write A Reference Letter
Articles about "how to write a reference letter"
http://jobsearch.about.com/lr/how_to_write_a_reference_letter/308151/2/

How To Write A College Recommendation Letter: If you are a high school teacher, you're probably inundated with requests for letters of recommendation ...
http://www.ehow.com/how_172521_write-college-recommendation.html

Reference Letter: Personal Letters of Recommendation: A recommendation letter is a way to build goodwill; reference letters from employers can be useful ...
http://www.buzzle.com/articles/reference-letters/

How To Write A Reference Letter
Wondering how to write a reference letter for employment or college admission?
http://answers.yourdictionary.com/language/how-to-write-a-reference-letter.html

Sample Business Letters of Reference
Sample of letter of reference in English. For ESL learners.
http://www.englishclub.com/business-english/correspondence-reference.htm

Writing A Reference Letter (with examples): At some point in your life , you're almost certainly going to have to write a reference letter for someone ...
http://www.dailywritingtips.com/writing-a-reference-letter-with-examples/

Recommendation Letter Sample Template
Writing a recommendation letter or letter of recommendation can be difficult ...
http://www.writinghelp-central.com/recommendation-letter.html

Reference Letter Definition and Samples: Reference letter sample templates are excellent to get you started when you need to write character reference letters ...
http://www.writinghelp-central.com/reference-letter-definition.html

How To Write A Recommendation Letter: A recommendation letter ideally starts by stating the name of the professor who writes the letter and his/her title ...
http://www.eastchance.com/howto/rec_let.asp

Sample Letters of Recommendation: Recommendation letter generator to create a reference letter for employment or a college letter of recommendation ...
http://www.boxfreeconcepts.com/reco/sample.html

Graduate School Letters of Recommendation: Most college graduate programs and many job opportunities require letters of recommendation ...
http://www.kelloggforum.org/graduate-school-letters-of-recommendation-recommendation-letter/

Recommendation Letter How-To Article: Article outlining how to write an effective recommendation letter, including what it should contain and omit ...
http://www.1st-writer.com/Recommendation.htm

College Admissions Recommendation Letter
College Admission Recommendation Letter for a disadvantaged student ...
http://www.lettersofrecommendation.net/

GENERAL WRITING REFERENCES

Even though this guide is all about "writing recommendation letters", as a Bonus I have included my standard quick-list of business writing resources for your information and reference, should you need additional writing help.

There are literally thousands of business writing reference books available. The following is a researched "desert island short-list" of what I consider to be some of the most accessible and useful books for anyone looking for basic help in drafting their business and/or general writing projects.

Writing Style References

A Pocket Style Manual, by Diana Hacker, Bedford/St. Martin's, 5th Edition (June 2009).
http://www.amazon.com

Basic Grammar in Use: Student's Book With Answers: Self-study Reference and Practice for Students of North American English, by Raymond Murphy with William R. Smalzer, (Sept. 2010).
http://www.amazon.com

The Chicago Manual of Style: The Essential Guide for Writers, Editors, and Publishers (16th Edition), by Chicago Press Staff, University of Chicago Press, 16th edition (August 2010).
http://www.amazon.com

The Economist Style Guide: The Bestselling Guide To English Usage, by The Economist Books (May 2010).
http://www.amazon.com

Effective Business Writing: A Guide for Those Who Write on the Job, by Maryann V. Piotrowski, HarperCollins,2nd Revised & updated edition (March 1996).
http://www.amazon.com

The Elements of Business Writing: A Guide to Writing Clear, Concise Letters, Memos, Reports, Proposals, and Other Business Documents, by Gary Blake, Robert W. Bly, Longman; 1st edition (August 1992).
http://www.amazon.com

The Elements of Style: The Original Edition, by William Strunk (August 2009).
http://www.amazon.com

How to Say It – Third Edition: Choice Words, Phrases, Sentences & Paragraphs for Every Situation (Paperback), by Rosalie Maggio (April 2009).
http://www.amazon.com

How to Write It: A Complete Guide to Everything You'll Ever Write, by Sandra E. Lamb, (Paperback) (July 2006).
http://www.amazon.com

Merriam Webster's Collegiate Dictionary, by Merriam-Webster, Merriam Webster, 11th Edition (Hardcover, April 2008).
http://www.amazon.com

MLA Handbook for Writers of Research Papers (7th Edition) , by Joseph Gibaldi, Modern Language Association of America, (March 2009).
http://www.mla.org

Publication Manual of the American Psychological Association, Sixth Edition, American Psychological Association, (July 2009).
http://www.apa.org

Professional Writing Skills, by Janis Fisher Chan, Diane Lutovich, Advanced Communication Designs, Inc.; 2nd edition (June 1997).
http://www.amazon.com

Writers Inc.: A Student Handbook for Writing and Learning, by Patrick Sebranek, , Dave Kemper, Verne Meyer, and Chris Krenzke., (Hardcover, Jan 2006).
http://www.amazon.com

Writing That Works: Communicating Effectively on the Job, by Walter E. Oliu, Charles T. Brusaw, and Gerald J. Alred (Paperback - Oct 2009).
http://www.amazon.com

Writing Tools

A collection of additional writing-related resources can be found at the following links at writinghelp-central.com:

Letter Writing (personal and business)
http://www.writinghelp-central.com/letter-writing.html

Business Writing
http://www.writinghelp-central.com/business-reports.html

APA Format
http://www.writinghelp-central.com/apa.html

MLA Format
http://www.writinghelp-central.com/mla.html

Writing Help Tools

In addition, my Writing Help Tools website also contains a number of other writing-related resources that may interest you:
http://www.writinghelptools.com

Other Writing Tools

The following is a list of my other writing-related toolkit ebooks:

***Instant Home Writing Kit* -**
How To Save Money, Time and Effort and Simplify Everyday Writing Tasks
http://instanthomewritingkit.com

***Instant Business Letter Kit* -**
How To Write Business Letters That Get The Job Done
http://instantbusinessletterkit.com

***Instant Recommendation Letter Kit* -**
How To Write Winning Letters of Recommendation
http://instantrecommendationletterkit.com

Instant Resignation Letter Kit -
How To Write A Super Resignation Letter and Move On With Class
http://instantresignationletterkit.com

Instant Letter Writing Kit –
How To Write Any Kind Of Letter Like A Pro
http://instantletterwritingkit.com

Instant College Admission Essay Kit -
How To Write A Personal Statement Essay That Will Get You In
http://instantcollegeadmissionessay.com

Instant Book Writing Kit -
How To Write, Publish and Market Your Own Money Making Book/eBook Online
http://instantbookwritingkit.com

How To Write A How-To Book (or eBook)
Make Money Writing About Your Favorite Hobby, Interest or Activity
http://howtowritehowto.com

INDEX

A

administrative assistant, 76

administrators, 4, 5, 16, 82

admission essay, 8, 195-198, 202-205, 216

Admissions Committee, 199

alexa.com, 217, 218

amazon.com, 7, 220, 221

appointment letter, 110

assessment statements, 55

attributes, 16, 55, 64, 74, 76, 84, 116, 124, 143, 155

award, 105, 144, 174, 181

B

bank reference, 163

barnesandnoble.com, 7

business, 161

business letters, 27, 28, 33

business undergrad, 146

business writing, 220

C

character reference letters, 17, 115, 116

characteristics, 17, 55, 64, 84, 90, 115, 116, 196-201

client letter, 156

closing statements, 59

colleague (reference), 154

college (admission), 1-8, 12-16, 44, 47, 63, 82-87, 91, 92, 116, 141, 177, 195-202, 205-207, 210, 216

commendation letters, 6, 7, 17, 167, 170

committee, 45, 47, 106, 111, 116, 144, 145, 158, 175, 183-187, 191, 197, 199

community, community service, 3, 5, 111, 157, 158, 162, 175

consultant, 97, 124, 134

corporate, 112

counseling, 111, 162

credit reference, 164

customer reference, 35, 161, 163

customer service, 132, 172

D

dental, 108

discriminatory, 66, 85, 119

domestic services, 155

downloadable templates, 5, 33, 38

E

emergency services (see medical)

employee review, 17, 177

employment related, 3

engineer, 127

exchange program, 145

ex-convict, 159

explain departure, 140

F

faculty members, 16, 82, 107, 201

fashion job recommendation, 79

fill-in-the-blank, 6, 33, 34, 35

financial, 76, 93, 130-136, 156

financial specialist (controller), 130,

friend (reference letter), 153

fully-formatted, 5-8, 33-39, 87, 180

G

general surgery (see medical)

government employees, 4

graduate program(s), 6, 16, 52-57, 82,
91-94, 106, 142

grammar, 26, 220

graphic artist (design), 69, 73

H

Haas (business school), 97

high school, 4, 44-47, 82, 88, 91, 136-
139, 143, 209-214

I

immigration letter (see visa)

insurance, 121, 158

internship, 80

IT program, 97,99, 134

IT specialist, 129

J

journalism, 101

L

landlord letter, 165

leadership, 44, 46, 55, 57, 65, 73, 77,
88, 93-95, 99, 107, 111, 122

letter generators, 39

letters of recommendation, 2, 3, 6, 8,
13, 15, 16, 19, 23, 63, 82, 92

letters of reference, 6, 12, 15, 63

letter-writing (tips), 3, 5, 6, 19, 39, 50,
222

M

manager, 78

marketing, 68, 81, 122, 126

master, 101

mature student, 89

MBA, 54, 95-97

medical residency...

- Emergency Services, 147

- General Surgery, 192

- Neurosurgery, 150

- Plastics, 148

MIS manager, 124, 134

N

neutral words, 66, 85

O

occupational therapy. 71

online resources, 18, 217

opening statement(s), 51

P

page compression, 29

parent letter, 173

performance evaluation letters, 1, 3,
15, 18, 63, 177-181, 185

pet owner, 165

Ph.D., 54, 90-95, 104, 107, 142, 145,
182, 185, 188-191

physician, 70, 147

privacy, 37, 67, 109, 160

project coordinator, 74

psychology, 85, 91, 104-107, 125

R

real-life templates, 4, 7, 18, 28, 31,
    33-42, 48, 50, 67, 109, 120, 141,
    152, 160
recommendation letters, 6, 7, 11, 12,
    15, 19, 23, 43, 64, 67, 82, 87, 222
reference letters, 1, 6, 7, 12, 42, 43, 63,
    109, 115, 117, 120, 141, 152, 160
request for letter, 191
research, 8, 11, 44, 46, 73, 84, 88, 91,
    92, 107, 126, 130-135, 142, 144,
    168, 181, 184-190, 195, 212-218
research specialist, 133
resignation letter, 34, 35, 44, 48, 223

S

sales, 1, 68, 97, 98, 121-123
sample letters (templates), 5-9, 33, 37,
    43, 67
satisfactory evaluation, 181
scholarship, 106, 107, 143, 187
security services, 136
software support, 128
special award, 113
Stanford, 95, 96
strategic plan, 49
student(s), 77, 89, 125, 138, 139, 145,
style guide, manual, 5, 6, 9, 31, 220

T

teacher(s), 4, 16, 44, 46, 70, 88, 107,
    137-139, 173, 189
teaching, 46, 47, 89, 106, 136-139,
    143, 144, 173, 178, 181-191

template adaptation method, 9, 38,
    41-45, 49, 50, 61
templates, 1-7, 27, 30, 31, 33, 36-39,
    41, 49, 51, 61, 64, 87, 170, 180
tenant letter, 166
transition words/phrases, 25
travel (see visa)

U

UK, 102, 103
undergraduate, 88, 90, 107, 143
university admission, 1, 3, 195
unsatisfactory evaluation, 189
unsolicited letters, 17, 167

V

visa (travel), 114
volunteer, 175

W

work performance, 16, 17, 64, 81, 177
writer's block, 41, 42
writing kits, 33
writing references, 7, 28, 220
writing tools, 222
writinghelp-central.com, 2, 3, 15, 23,
    222

NOTES

NOTES

NOTES

NOTES

Made in the USA
Lexington, KY
15 March 2015